INTERRELATIONS BETWEEN PEOPLE AND PETS

INTERRELATIONS BETWEEN PEOPLE AND PETS

Edited by

BRUCE FOGLE, D.V.M., M.R.C.V.S.

Chairman and Editor, Symposium on the
Human–Companion Animal Bond
London, England

With a Foreword by

ANDREW EDNEY

President
British Small Animal Veterinary Association

CHARLES C THOMAS • PUBLISHER
Springfield • Illinois • U.S.A.

Published and Distributed Throughout the World by

CHARLES C THOMAS • PUBLISHER
Bannerstone House
301-327 East Lawrence Avenue, Springfield, Illinois, U.S.A.

© *1981, by* CHARLES C THOMAS • PUBLISHER

ISBN 0-398-04169-5

Library of Congress Catalog Card Number: 80-22692

*With THOMAS BOOKS careful attention is given to all details of manufacturing
and design. It is the Publisher's desire to present books that are satisfactory as to their
physical qualities and artistic possibilities and appropriate for their particular use.
THOMAS BOOKS will be true to those laws of quality that assure a good name
and good will.*

Library of Congress Cataloging in Publication Data
Main entry under title:

Interrelations between people and pets.

    Bibliography: P.
    Includes index.
    1. Pets — Social aspects — Congresses.   2. Pet
owners — Psychology — Congresses.   3. Pets — Behavior —
Congresses.   I. Fogle, Bruce.
SF411.5.I57   636.08'87   80-22692
ISBN 0-398-04169-5

Printed in the United States of America

S-RX-1

CONTRIBUTORS

ALAN BECK, Sc.D. Director, Center on the Interactions of Animals and Society, University of Pennsylvania, Philadelphia, Pennsylvania

LEO BUSTAD, D.V.M. Dean and Professor, College of Veterinary Medicine, Washington State University, Pullman, Washington

JULES CASS, D.V.M., M.S. Chief Veterinary Medical Officer, Department of Medicine and Surgery, Veterans Administration, Washington, D.C.

SAMUEL CORSON, B.S., M.S., Ph.D. Professor, Department of Psychiatry and Academic Faculty of Early and Middle Childhood Education, Ohio State University, Columbus, Ohio

ELIZABETH O'LEARY CORSON, B.S., M.S. Research Associate, Department of Psychiatry and Academic Faculty of Early and Middle Childhood Education, Ohio State University, Columbus, Ohio

BRUCE FOGLE, D.V.M., M.R.C.V.S. Practicing Veterinarian, Chairman and Editor, Symposium on the Human-Companion Animal Bond, London, England

MICHAEL FOX, D.Sc., Ph.D., B. Vet. Med., M.R.C.V.S. Director, Institute for the Study of Animal Problems, Washington, D.C.

GISELHER GUTTMANN, Professor, Institute of Psychology, University of Vienna, Vienna, Austria

LINDA HINES, Research Associate, College of Veterinary Medicine, Washington State University, Pullman, Washington

AARON KATCHER, M.D. Professor, Department of Psychiatry, University of Pennsylvania, Philadelphia, Pennsylvania

MICHAEL McCULLOCH, M.D. Associate Clinical Professor of Psychiatry in Veterinary Medicine, Oregon State University, Portland, Oregon

ALASDAIR MACDONALD, M.Bc.Hb., M.R.C. Psych., D.P.M., D.C.H. Lecturer in Psychiatry, Ninewells Hospital and Medical School, Dundee, Scotland

v

PETER MESSENT, M.A., D. Phil., Animal Behaviorist, Animal Studies Centre, Waltham-on-the-Wolds, Leicestershire, England

ROGER MUGFORD, B.Sc., Ph.D., Clinical Animal Behaviorist, Melton Mowbray, Leicestershire, England

CONSTANCE PERIN, Ph.D., Cultural Anthropologist, Cambridge, Massachusetts

MARIALISA ROMASCO, M.S.W. Resident Social Worker, Companion Animal Clinic, School of Veterinary Medicine, University of Pennsylvania, Philadelphia, Pennsylvania

ELEANOR RYDER, M.A., M.S.W. Professor, School of Social Work, University of Pennsylvania, Philadelphia, Pennsylvania

JAMES SERPELL, B.Sc., Ph.D. Zoologist, Subdepartment of Animal Behavior, University of Cambridge, Cambridge, England

JAMES VAN LEEUWEN, M.D. Senior Staff Psychiatrist, Hospital for Sick Children, Toronto, Ontario, Canada

VICTORIA VOITH, D.V.M., M.Sc., M.A. Associate in Medicine, School of Veterinary Medicine, University of Pennsylvania, Philadelphia, Pennsylvania

CEILE WASHINGTON, D.V.M. Resident Veterinarian, Companion Animal Clinic, School of Veterinary Medicine, University of Pennsylvania, Philadelphia, Pennsylvania

ANDREW YOXALL, M.A., Vet. M.B., M.R.C.V.S. Lecturer, School of Veterinary Medicine, University of Cambridge, Cambridge, England

FOREWORD

From time to time there is a need to gather in knowledge from a variety of seemingly unrelated fields and to present it in an integrated manner. The British Small Animal Veterinary Association felt that there was a need to do this in the field of pet–owner relations and in January 1980, held an international symposium on The Human–Companion Animal Bond.

The B.S.A.V.A. membership is made up of veterinarians who have an interest in small animals. What are called "companion animals" mostly come under the heading of small animals. This work makes up around 50 percent of all veterinary practice in the United Kingdom.

It is obvious to all veterinarians who have spent any time in small animal practice that pet animals may bring many benefits to their owners and families. What most of us would find very difficult to do would be to actually evaluate and quantify such benefits. This book does just that.

Andrew Edney, President
British Small Animal Veterinary Association

PREFACE

Almost half of the homes of western Europe and North America have pets. And although caged and aquaria pets are inexpensive to buy and maintain; although the cat makes the ideal pet for the urban dweller, most pet owners keep dogs.

Why do so many people burden themselves so? What are their motives? What are their rewards?

This book answers some of these questions. It contains papers presented at the Human–Companion Animal Bond Symposium, sponsored by the British Small Animal Veterinary Association, held in London in early 1980, supplemented by several additional chapters discussing bonding as it pertains to veterinary practice.

In Section I, writers describe the development of the pet-owner bond from psychological, sociological, ethological, anthropological, and psychiatric perspectives.

Section II is devoted to a pragmatic discussion of possible therapeutic uses of pets.

Section III describes how veterinary medicine can be integrated more closely to the social sciences. With this in mind, guidelines for urban planning and veterinary curricula are presented.

Section IV discusses problems that arise from the human-companion animal bond; problems of communication, problems of obedience training of pets, and problems involving euthanasia and grieving.

This book is a beginning toward a future interdisciplinary understanding of why we keep pets. Through understanding this bond, we will better understand outselves.

Bruce Fogle

ACKNOWLEDGMENTS

The British Small Animal Veterinary Association wishes to acknowledge the generous assistance of the following in supporting the symposium.

Carnation Foods Company Limited
The Latham Foundation
Merck Sharp & Dohme Limited
Pedigree Petfoods
SmithKline Animal Health Limited
Spillers Limited
Tetra (Warner Lambert UK Limited)

The Association is grateful to the following for their help in typing of the symposium proceedings.

A.H. Robins Limited
Beecham Animal Health Limited
The Boehringer Corporation Limited
DVM Pharmaceuticals Limited
Glaxovet Limited
Hoescht Pharmaceuticals Limited
Mycofarm Limited
Pedigree Petfoods
SmithKline Animal Health Limited
The Veterinary Drug Company Limited

The Association would like to thank the following for having served on the organizing committee for the symposium.

Bruce Fogle (Chairman)   Colin Price
Meredith Lloyd-Evans   John Sampson
John Oliver   Tricia Parker
Michael Keeling

CONTENTS

                                                                    Page

Foreword .... *ix*

Preface .... *xi*

SECTION I.
THE DEVELOPMENT OF THE
HUMAN-COMPANION ANIMAL BOND

Chapter

1. A Historical and Biological View of the Pet–Owner
   Bond
   Peter Messent and James Serpell 5
2. Relationships Between the Human and Non-human
   Animals
   Michael Fox 23
3. Interactions Between People and Their Pets:
   Form and Function
   Aaron Katcher 41
4. Dogs as Symbols in Human Development
   Constance Perin 68
5. The Psychological Determinants of Keeping Pets
   Giselher Guttmann 89

SECTION II.
MEDICAL APPLICATIONS OF THE
HUMAN-COMPANION ANIMAL BOND

6. The Pet as Prosthesis: Defining Criteria for the
   Adjunctive Use of Companion Animals in the
   Treatment of Medically Ill, Depressed Outpatients
   Michael McCulloch 101
7. Pet Facilitated Therapy in Human Health Care
   Jules Cass 124

8.   Companion Animals as Bonding Catalysts in
     Geriatric Institutions
     Samuel Corson and Elizabeth O'Leary Corson 146
9.   A Child Psychiatrist's Perspective on Children and
     Their Companion Animals
     James VanLeeuwen 175
10. The Pet Dog in the Home: A Study of Interactions
     Alasdair Macdonald 195

SECTION III.
THE HUMAN–COMPANION ANIMAL BOND IN
SOCIAL PLANNING AND IN VETERINARY SCHOOLS

11. Establishing a Social Work Service in a Veterinary
     Hospital
     Eleanor Ryder and Marialisa Romasco 209
12. The First Year of an Animal Behavior Clinic in a
     Veterinary School: Process and Outcome
     Ceile Washington 221
13. Guidelines for Planning for Pets in Urban Areas
     Alan Beck 231
14. A Curriculum to Promote Greater Understanding
     of the Human–Companion Animal Bond
     Leo Bustad and Linda Hines 241

SECTION IV.
VETERINARY PRACTICE AND THE
HUMAN–COMPANION ANIMAL BOND

15. Attachment Between People and Their Pets:
     Behavior Problems of Pets that Arise from the
     Relationship Between Pets and People
     Victoria Voith 271
16. Problem Dogs and Problem Owners: The Behavior
     Specialist as an Adjunct to Veterinary Practice
     Roger Mugford 295

17. Client Problems as Presented to the Practicing
Veterinarian
Andrew Yoxall.....................................318
18. Attachment — Euthanasia — Grieving
Bruce Fogle.......................................331

Index...345

INTERRELATIONS BETWEEN PEOPLE AND PETS

SECTION I.

THE DEVELOPMENT OF
THE HUMAN–COMPANION
ANIMAL BOND

CHAPTER 1

AN HISTORICAL AND BIOLOGICAL VIEW OF THE PET–OWNER BOND

PETER R. MESSENT AND JAMES A. SERPELL

Two species of companion animal—the dog and cat—stand out in terms of their popularity from all other species kept as domestic pets. It has often been suggested that the wild ancestors of dogs and cats were strongly preadapted for life in a combined social group with early man, and in this discussion we examine the nature of these suggested preadaptations. Also, through a general review of the historical, behavioral, and ecological background to man's lengthy association with these two species, we hope to shed further light on the biological and cultural origins of the human–companion animal bond.

TIME AND PLACE OF DOMESTICATION

The dog is generally thought to be the oldest of domestic animal species (Zeuner, 1963; Scott, 1968). The earliest fossil remains, dating from about 12,000 years ago, come from Iraq and Israel (Turnbull and Reed, 1974; Davis and Valla, 1978). Later finds have come to light in the United States from about 10,000 years ago (Lawrence, 1967), Denmark and the United Kingdom about 9,000 years ago (Degerbøl, 1961; Musil, 1970), and from China about 7,000 years ago (Olsen and Olsen, 1977). The available evidence therefore suggests that the earliest domestication of the dog took place in the Near East during the preagricultural Mesolithic period of human cultural development, which followed the end of the last global iceage.

This period is characterized in the archaeological record by a shift from the seminomadic and specialized hunting economies of the ice age to the more mixed hunter-gatherer economies and relatively settled communities that arose following the retreat of the

5

ice caps (Harris, 1969). The warmer and wetter climate resulted in a rich and diverse flora and fauna, particularly in the river basins and coastal plains, and the communities that settled there ultimately gave rise to complex agricultural civilizations, as in the Indus Valley, Sumer, and Egypt.

The origins of the domestic cat are less easy to trace than those of the dog, mainly because its skeletal morphology differs only slightly from that of its presumed wild ancestor, the *lybica* group of *Felis sylvestris*, which occurs throughout North Africa and the Near East (Todd, 1978). Zeuner (1963) has postulated a date of about 9,000 years ago for the earliest domestication of the cat, but as yet there is no fossil evidence to support this view. Certain authors, e.g. Fox, 1975, have assumed a connection between the domestication of the cat and the control of rodent pests. If this assumption is correct, then the domestication may have occurred much later, since granivorous rodents are unlikely to have become serious pests before the emergence of large agrarian societies around 4,000 B.C. Representations of domestic cats do not appear in the iconography of ancient Egypt before the third millenium BC (Smith, 1969).

In many ways, the knowledge of when a species was first domesticated is only of limited usefulness in a discussion of the origins of the human–companion animal bond. Domestication can be defined as that point at which the care, feeding and, above all, the breeding of a species comes under the control of man (Kretchmer and Fox, 1975). In other words, the point at which man, as opposed to natural forces, becomes the primary selective agent in a species' evolution. The date of domestication tells us only when the effects of man's selective breeding first become manifest and provides no clues as to how long a close association may have existed prior to this.

When European explorers first set out to discover the uncharted regions of the world, they encountered innumerable preagricultural village-communities overrun with pets of every description: wild animals captured and raised in captivity as companions (Sauer, 1952). It seems highly probable that pet keeping arose as a nearly universal human proclivity long before actual domestication took place. Recent fossil discoveries from China, indicating a possible association between Peking man and a wolflike canid as early as 500,000 years ago offer tentative support for this hypothesis (Olsen and Olsen, 1977). However, one major question that this raises is what social and/or environmental factors led to the switch from the mere capture and taming of wild animals as pets to the process of true domestication?

THEORIES OF DOMESTICATION
Domestication for food

While this was undoubtedly the stimulus for the domestication of the major meat-producing artiodactyl species, such as sheep, goats, and cattle, few authors support the possibility that dogs and cats might have been domesticated solely as items of food. It is true that, unlike the cat, the dog was and still is eaten in certain parts of the world, notably the Far East and the Americas, although the practice is nowhere widespread (Degerbøl, 1961; Zeuner, 1963). Also, it has been suggested that young wild canids, captured for "the pot" during times of food shortage, might have escaped this fate, been tamed, and thus gave rise to domestic stock (Fuller and Fox, 1969). However, dogs and cats are predominantly carnivorous, and this makes them relatively poor propositions as food species by comparison with herbivorous primary consumers, such as ruminants. It therefore seems likely that the practice of dog eating originally arose as a temporary starvation measure.

Domestication as Guards

The cat is unsuitable for guarding or warning because it lacks the necessary behavior required to reliably alert people to danger. By contrast, the domestic dog exhibits territorial barking at much higher frequencies than is reported in its postulated wild ancestor the wolf (Fox, 1971). This would imply that hypertrophied warning or guarding behavior has been deliberately selected by man. Undoubtedly, dogs as early warning devices would have been useful additions to any Mesolithic settlement, given man's warlike tendencies. However, although it was a valuable by-product, it is difficult to imagine how territorial barking on its own could have provided the original stimulus for domestication.

Domestication for Hunting

Pictorial representations from the third millenium BC in ancient Egypt illustrate cats being used for hunting and retrieving (Smith, 1969). The practice evidently disappeared by the second millenium BC; it is unlikely to have ever been an important functional role for cats in society.

The tendency to hunt in cooperative packs, found in several wild canids, has led to the theory that man's earliest association with wild dogs or wolves took the form of an opportunistic exploitation by man of the wild dogs' hunting skills.

Downs (1960) postulates that human hunting parties, recognizing the superior tracking abilities of wild dogs, might have followed dog packs to the kill and then deprived them of the carcass, possibly leaving the offal for the dogs. There is evidence of a similar exploitation of wild dingoes in Australia by aboriginal hunting parties (Meggit, 1965). This initially parasitic relationship might have, in time, evolved into a more symbiotic one in which both species could have derived mutual benefit.

By contrast, Sauer (1952) feels that the idea of an early hunting symbiosis between men and dogs is a myth propagated by modern European romantics. He points out that hunting dogs are late specializations among certain peoples of advanced culture and that truly primitive people do not use dogs for hunting, although they may accompany the hunting party on occasions.

Meggit's (1965) observations of Australian aborigines would tend to support this conclusion. Aborigines raise dogs and sometimes wild dingo cubs and treat them with great affection. However, they do not regard them as aids to the hunt, and the dogs themselves show little inclination to participate.

Certainly, modern hunting dogs are generally modified in some way as to make them more appropriate to man's normal hunting methods. Where they are encouraged to hunt in packs like wolves, the dogs tend either to have short legs and a slow running speed so that men can keep up with them on foot, or they are pursued on horseback (Scott, 1968).

Overall, this would indicate that the hunting methods of wolves and men were originally incompatible and could not have given rise to an early symbiosis or to domestication.

Domestication via Commensalism

As stated earlier, it is popularly held that the original association between men and cats arose as a form of commensalism in which wild cats invaded human agricultural settlements in order to feed off refuse and steadily increasing populations of rodent pests, and were actively encouraged to do so by their human hosts (Fox, 1975; Todd, 1978).

It has often been suggested that a similar association resulted in the domestication of the dog. Zeuner (1963) felt that tolerance by man of scavenging wild dogs, feeding off the abandoned refuse of his hunting camps, permitted the gradual development of a progressively more intimate relationship between the two species which eventually led to a permanent one.

Downs (1960) elaborates this idea and offers an explanation for why the domestication took place when it did. He suggests that in the Mesolithic period the way of life of human groups changed from a more or less nomadic existence to one involving more permanent settlements. Thus, wild canids would have changed from being occasional campfollowers to being permanent camp scavengers. The scavenging role of hand reared dogs around the camp sites of Australian aborigines lends some support to this theory (Montagu, 1942).

Zeuner (1963) regards the pariah dog—the ubiquitous canine scavenger of North Africa and Asia—as a surviving relic of an early commensal association between men and wild dogs. However, Hilzheimer (1932) concludes that the pariah merely represents the surviving remains of truly domestic dogs made homeless by past economic and cultural upheavals.

The major objection to the commensal theory of domestication is that during the Mesolithic period when the domestication is thought to have occurred, human populations were probably too small and produced too little refuse to provide waste in sufficient quantities to sustain a permanent population of scavenging wolves. Modern pariah dogs occur characteristically in areas of very high population density and poor or nonexistent means of waste disposal.

Domestication as Pets

One of the most popular theories of the domestication of the dog is entirely free of utilitarian considerations. Several authors postulate that the earliest domestication arose as a natural and, in some ways, inevitable consequence of the universal human tendency to adopt young wild animals as pets (Sauer, 1952; Reed, 1954; Scott, 1968). These authors picture young wolf cubs being brought into the villages of early man by returning hunters or children and then being adopted and even suckled by human females. Under these circumstances, the animal might acquire a personal name and a status equal to that of other members of the family group.

The popularly held view that wolves are innately savage and untameable is clearly an erroneous one, since several contemporary accounts exist of hand reared and even adult wolves becoming as tame and affectionate toward humans as domestic dogs (Crisler, 1958; Fentress, 1967; Woolpy and Ginsberg, 1967).

The problem with this theory is that it fails to explain why pet keeping led to domestication when it did and not before. As Downs (1960) points out, while pet keeping is virtually ubiquitous among all human groups, the phenomenon of domestication is strictly localized in its early stages and, under the circumstances, one would expect the two processes to be coincident in all areas. However, fossil discoveries already described indicate that once domestication was underway, domestic dogs spread rapidly throughout human societies all over the world, and one could equally argue that this would have rendered further domestication attempts unnecessary.

It is apparent from the archaeological record that the Mesolithic period in the Near East involved fundamental changes in human culture. With the retreat of the ice caps, the comparatively meager subsistence economies of the Palaeolithic period gave way to a much richer and more settled way of life (Harris, 1969). The significant rise in standards of living would have created more leisure time for humans to indulge in essentially nonproductive activities such as art, religious ritual, and simple experimentation in the taming and cultivation of wild animals and plants.

Above all, these societies probably had long-term surpluses of food, and this would have been a vital factor enabling them to sustain captive breeding populations of wild animals such as dogs.

With insufficient food, few wild species will breed successfully in captivity, and as stated earlier, some form of captive breeding is an essential prerequisite for domestication to occur. Surviving Australian aborigines are notable for having an economy that barely approaches the subsistence level (Pfeiffer, 1972). One result of this is that their hand reared dingoes, while treated with great affection, are so severely undernourished that they rarely if ever breed, and Meggit (1965) suggests this as a reason why a breed of domestic dingoes has not arisen in Australia. In view of this, it seems reasonable to postulate that pre-Mesolithic pets throughout the world were subject to comparable deprivation and could not therefore have been domesticated.

On the basis of the present evidence it appears likely that the human–companion animal bond is of great antiquity and that it was this bond or, more specifically, man's capacity to generalize his social responses to include wild animals of other species, that lies at the roots of the domestication process. At least with regard to the dog, useful canine attributes, such as scavenging, hunting, and territorial barking, may have helped to cement the attachment but were probably not essential to its initial formation.

The fundamental switch from simple pet keeping to true domestication was probably precipitated by relatively sudden and favorable changes in the environment in certain localized geographic regions. Until further archaeological evidence is available, one can only speculate as to whether cat domestication proceeded along similar lines. However, the fact that the cat is the second most popular companion animal in modern societies, irrespective of its rodent catching habits, suggests that this may have been the case.

WHY DOGS AND CATS?

A broad consideration of the principal factors that have contributed to the outstanding success and popularity of dogs and cats as pets can provide further insights into the nature of the human-companion animal bond. Why, for example, were the Asian subspecies of wolves and wildcats, rather than other available wild species, the only ones initially chosen as the subjects for domestication? One obvious explanation is that once the role of "companion animal" had been filled by dogs and cats, there was no longer any need for men to adopt alternative species.

Striking confirmation for this idea comes from studies of the Falkland Island wolf (now extinct) by Clutton-Brock (1977). Historical and osteological evidence suggests that the natives of the Falkland Islands, in the absence of the domestic dog, tamed and partially domesticated the indiginous, wolflike canid *Dusicyon australis*. The species was, however, displaced and became extinct following the introduction of European domestic dogs. Clearly, *Canis familiaris* was better adapted to the role of companion than the native species. However, one is still left wondering what, if anything, originally gave wolves and wild cats "the edge" over other species for the exclusive status of companion animal. Several authors, e.g. Reed, 1954; Herre, 1963; Scott, 1968, have argued that these species, especially the wolf, were already strongly preadapted in various ways to life as members of human social groups.

DOG AND CAT PREADAPTATIONS
Primary Socialization

The majority of young mammals, if obtained at the appropriate age, can be tamed easily and "socialized" to humans or to other species. This capacity to rapidly form social attachments is evidently confined to a relatively short "sensitive period" during early development (*see* Bateson, 1979), after which positive social responses to strangers gradually decline and are replaced by fearful behavior that effectively prevents the formation of further attachments. This "primary socialization period" (Scott, 1963) varies considerably in its onset and duration from species to species.

For example, in the majority of ungulates, in which the young are born in a precocial state, primary socialization appears to occur within a few hours of birth, whereas in dogs and cats it begins at around three weeks of age (Fox, 1975; Scott, 1963), corresponding to the time when the young first begin exploring outside the dark interior of the den or nest. If puppies or kittens are exposed to individuals of two (or more) species during the few weeks of socialization, they will socialize with both (Fox, 1967).

From the point of view of domestication, the most important feature of the socialization experience is that it determines the species and, in some cases, the individuals to which a young mammal will respond in a positive social and sexual manner when adult. In other words, if a wolf cub is entirely hand reared by humans throughout the socialization period, it will subsequently fail to recognise wolves as members of its own species and consequently would be useless for breeding or domestication. However, if obtained from six to eight weeks of age, it will already have had time to socialize with its parents and littermates and will still be capable of establishing further attachments with humans. As an adult, such an individual would be able to breed successfully while still owing its primary allegiance to humans.

Clearly, the existence of a relatively late and protracted primary socialization period in dogs and cats will have facilitated their early domestication as companions, although this is an attribute that they share with many other carnivore species.

Social Organization and Territory

The wolf is one of the most highly social species of canid, and it specializes in communal living associated with the cooperative hunting of large game animals (Kleiman, 1967). Wolf packs usually comprise between ten and twenty genetically related individuals, and the social structure is hierarchically organized by means of strictly maintained dominance-subordinance relationships, controlled by a pair of dominant ("alpha") animals of each sex (Woolpy, 1968; Fox, 1971). The pack possesses a home range or hunting territory from which strange individuals are actively excluded (Joslin, 1967).

In many respects, this type of social organization is similar to that postulated to exist among prehistoric hunting cultures of man (Etkin, 1962). Among the thirty-four other species of living canid, only the Dhole (*Cuon alpinus*) and the African hunting dog (*Lycaon pictus*) are known to be as highly social, and this suggests another reason why wolves were prime contenders for the position of social partner with early man. It should also be stated that neither of these two other species are recorded from the Mesolithic

Near East. The tendency of domestic dogs to acknowledge the dominant status of humans is undoubtedly an ancestral characteristic derived from pack life. It is probably indispensible for the maintenance of good relations between men and dogs.

In contrast to the dog, the cat has never been considered a particularly social species, although recent studies have shown that they are naturally more sociable than was previously suspected (Macdonald and Apps, 1978; Fagen, 1978). Without doubt, the majority of cat owners would agree that cats can be very sociable and affectionate while still maintaining a degree of independence that is uncharacteristic of dogs. Cats are also highly territorial (Fox, 1975), and this is another factor that tends to tie them to the human habitation and its inmates.

Communication

Dogs and cats have only a limited repertoire of vocalizations and certainly nothing comparable to human language. Nevertheless, the ability of these pets to communicate with their owners by nonvocal means—or at least to make their owners believe that this is occurring—is, and probably always was, very important in the pet–owner relationship (*see* Chapter 3). Both the dog and cat employ body postures and various expressive behaviors that owners believe they can readily interpret, and both have relatively large repertoires of visual social signals compared with many wild species. Facial expressions, changes in the position of the body, ears, and tail or in the direction of gaze can all be used to express a wide range of different emotional states.

The wolf, in particular, shows a greater range and variability of expressions than any other species of canid (Fox, 1970). Bolwig (1962) states that primates (humans) and dogs (wolves) have similar facial expressions, and the facial muscles that they have in common are used in a similar manner when expressing emotion. In addition, the fact that the social signals of wolves and men had evolved for communication in similarly organized societies would have further enhanced the bond of similarity between the two species.

Play

The dog is reported to show more play as an adult than most other species of wild or domestic animal (Aldis, 1975). Clearly, this behaviour has been enhanced by artificial selection, although wolves also play quite often as adults (Mech, 1970). Cats, especially as kittens, also have a reputation for playfulness (Fox, 1975).

Play may be an important feature of the human–companion animal bond because it greatly increases the potential for contact between the pet and its owner. Above all, the noncompetitive qualities of pet orientated play provide an absorbing and harmless form of recreation for human beings of all ages, and the need for companion animals that reciprocated man's own playful tendencies may well have influenced his choice of species for domestication.

Intelligence

With the exception of primates, dogs and cats are among the most intelligent of terrestrial mammals, and it is an attribute that most pet owners admire. Relatively high intelligence has probably evolved in these species as an adaptation to hunting, and undoubtedly it helps them to adapt more readily to the idiosyncracies of life in partnership with man.

In view of the importance of intelligence, it might seem surprising that man has never domesticated more intelligent species, such as primates. There are several likely reasons for this. Among the smaller primates that might be suitable as pets, the majority are highly arboreal and very difficult to house train. Their complex social organizations involve exceptionally high levels of competitive aggression (Jolly, 1970), and this characteristic is all too easily transferred to humans, making even tame adult primates generally unpredictable and potentially dangerous. Also, their very intelligence means they demand more attention than a dog or cat from their owner or else they become frustrated and may display aberrant destructive behaviors. Primates would therefore need more attention than the vast majority of potential owners would be prepared to commit. It has recently been reported that the chimpanzee will display deceitful behavior in certain circumstances (Woodruff and Premack, 1979). Such deceitful or spiteful behavior, if ever displayed, would clearly by unsuitable in a domestic pet.

Activity Cycles

Humans are generally active by day, and this is the time when they wish to interact with their pets. The popularity of the dog may in part be due to the fact that it shares man's diurnal habits. Cats are crepuscular rather than truly nocturnal and are most active around dawn and dusk. This corresponds to the time when most people are at home and have plenty of time to interact with their pets.

It is probably significant that only recently, with the invention of electric lights, have nocturnal rodents such as rats, mice, and hamsters become popular as pets.

Elimination Behavior

Dogs and cats are easily house-trained, but this would not be the case with a great many wild animal species. Carnivores in general are selective and fairly predictable about where they leave their urine and feces, and this naturally tidy aspect of their behavior probably contributed greatly to their early success as house pets. It is significant that in other popular pets, such as rodents and cage birds, the problem of undesirable eliminative behavior is solved by keeping them in relatively closed environments (cages) where their urine and feces are isolated from the home environment.

While several herbivores have been domesticated, and some individuals are regarded as pets by their owners, these need to be housed outside. This greatly reduces the potential for touch and talking interactions that seem to be one crucial aspect of the pet–owner bond.

Physical Characteristics

Obviously, some species would be unsuitable as pets for purely physical reasons. Very large or very small size, specialized feeding habits, or the production of disagreeable odors may all be important factors in ruling out certain species. It is notable that comparatively small size in domestic dogs was one of the first characteristics to be selected by early man.

It is clear that dogs and cats were indeed strongly preadapted to the niche of companion animal in human society and that it was these preadaptations combined with coincidental factors, such as geographic distribution and the Mesolithic cultural revolution in the Near East, that singled them out as the subjects for domestication. Since domestication, both species have been modified in various ways by a process of conscious and unconscious selective breeding by man. An examination of the common trends in this process can be used to demonstrate the sorts of morphological and behavioral characters that man found desirable, undesirable, or unimportant in his relationship with his pets.

CHANGES RESULTING FROM DOMESTICATION

Morphological Changes

As stated earlier, the domestic cat differs surprisingly little from its supposed wild ancestor. Only the length and color of the coat show major modifications. Such changes may give cats greater tactile and visual appeal for their owners compared with the natural coat.

In contrast, the domestic dog exhibits truly outstanding variability in virtually every possible morphological character (Ginsburg, 1976). In part, this is due to the natural genetic flexibility of the dog (Scott and Fuller, 1965), although man has clearly favored selection of some characteristics over others. The vast majority of domestic breeds are smaller than the wild ancestor and consequently require less food.

Most dogs have sickle shaped or curly tails that are held higher than they would be in normal wolves, and in many breeds the ears are drooping instead of erect. Changes in the position of the ears and tail are vitally important for intraspecific communication in the wolf, while the structural changes to these organs in the dog have greatly limited their signalling capacity. Humans, of course, do not signal with their ears or tail; it appears likely that man has not considered the natural mobility of these organs important in his relationship with the dog and has relaxed selection in favor of its preservation.

Changes in coat color and length may be in the same category since piloerection is also important in wolf communication. Another common trend has been a gradual foreshortening of the head and muzzle, culminating in flat-faced breeds, such as the bulldog, pug, and Pekingese. Many breeds also have abnormally large and frontally directed eyes.

It has often been pointed out that the universally appealing nature of young animals is due to their round, flattish faces and relatively large eyes. The morphological trends in the breeding of small lap dogs strongly suggest that man has been selecting infantile or paedomorphic characteristics. Anthropomorphic and therefore comically appealing attributes are also widespread among domestic breeds.

Overall, it would appear that man has bred dogs primarily to be more like himself or his children and has thus brought them closer to his ideal concept of companion. Naturally, utilitarian considerations, such as size, speed, and strength for working or hunting, have also played a part in moulding the evolution of domestic dogs.

BEHAVIORAL CHANGES

Neither the dog nor the cat differs markedly in behavior from its wild ancestor. This is particularly surprising in the case of the dog, given its extraordinary morphological variability, and is further confirmation of the idea that this species' behavior was already strongly preadapted for domestic life (Scott, 1968). The most important behavioral modifications resulting from domestication are changes in response threshold to certain stimuli, increased docility and adaptability, perpetuation of infantile behavior patterns into adulthood, and a trend toward promiscuous rather than pair-bond mating (Kretchmer and Fox, 1975).

An example from the first of these categories would be the relatively exaggerated territorial barking of domestic dogs that has already been discussed. This behavior reflects a genetically based reduction in the levels of stimulation that will evoke territorial behavior. Increased docility and adaptability are also the products of genetic changes in response threshold (Scott and Fuller, 1965). These two factors are both essential adaptations to the role of companion animal and probably appeared early in the histories of dogs and cats. By selective breeding, Belaev and Trut (1975) were able to significantly increase tameness and docility in wild silver foxes within a few generations. Selection in favor of paedomorphic characters in domestic dogs has already been described in the previous section, and it is likely that the perpetuation of infantile behavior patterns in adult pet animals has occurred for similar reasons. Many owners talk to their pets and treat them as children, and it is likely that infantile behaviors on the part of the pet help to enhance this response. Such behaviors include play, passive submission (Schenkel, 1967), and whining in the dog, and kneading and possibly purring in the cat.

In nearly all breeds of domestic dog, bitches come in heat twice rather than once yearly, and male dogs are sexually active all year round. Unlike the wolf, dogs do not generally display complex courtship behavior prior to copulation. All these changes can be interpreted as opportunistic adaptations to domestic life.

CONCLUSIONS

The domestications of the cat and dog occurred when and where they did for a combination of reasons. Radical changes in human culture toward richer and more settled ways of life in certain areas of the Near East created a social and economic environment appropriate to domestication experiments. Man had probably captured and raised wild animals as pets long before this, but it was specifically the changes in "climate" of the Mesolithic period that allowed this early association to evolve into the stronger and more durable bond of domestic partnership. Not surprisingly, the species that first achieved the status of domestic companions were those, such as the dog and cat, that were already better adapted to fit this role. Above all, these species were able to transfer their normal social attachments to man and to behave toward him in a manner that he interpreted as friendly, affectionate, and companionable. Since domestication, the effects of man's selective breeding have been, in the main, to enhance the anthropomorphic and friendly qualities of these two species. In so doing, he has firmly cemented the bond between human and companion animal.

REFERENCES

Aldis, O.: *Play Fighting*. Academic Press, New York, 1975.

Bateson, P. P. G.: How do sensitive periods arise and what are they for? *Anim Behav, 27*:470–86, 1979.

Belaev, D. K. and Trut, L. N.: Some genetic and endocrine effects of selection for domestication in silver foxes. In Fox, M. W. (Ed.): *The Wild Canids*. Van Nostrand Reinhold, New York, pp. 416–26, 1975.

Bolwig, N.: Facial expression in primates with remarks on parallel development in certain carnivores. *Behaviour, 22*:167–92, 1962.

Brothwell, D. and Higgs, E.: *Science in Archaeology*. Thames and Hudson, London, 1963.

Clutton-Brock, J.: Man-made dogs. *Science, 197*:1340–42, 1977.

Crisler, L.: *Arctic wild*. Harper Bros., New York, 1958.

Davis, S. J. M. and Valla, F. R.: Evidence for domestication of the dog 12,000 years ago in the Natufian of Israel. *Nature, 276*:608–10, 1978.

Degerbøl, M.: On a find of a preboreal dog (*Canis familiaris L.*) from Star Carr, Yorkshire, with remarks on other Mesolithic dogs. *Prehistoric Soc for 1961 New Ser, 27*:35–55, 1961.

Downs, J. F.: Domestication: an examination of the changing social relationships between man and animals. *Kroeber Anthrop Soc Papers, 22*:18–67, 1960.

Etkin, W.: Social behaviour and the evolution of man's mental faculties. In Montagu, M. F. A. (Ed.): *Culture and the Evolution of Man*. Galaxy Books, London, 1962.

Fagen, R. M.: Population structure and social behaviour in the domestic cat (*Felis catus*). *Carnivore Genet Newsletter, 3*:276–81, 1978.

Fentress, J. C.: Observations on the behavioural development of a hand-reared male timber wolf. *Amer Zool, 7*:339–51, 1967.

Fox, M. W.: The effects of early experience on the development of inter- and intra-specific social relationships in the dog. *Anim Behav, 15*:377–386, 1967.

_____: A comparative study of the development of facial expressions in canids, wolf, coyote and foxes. *Behaviour, 36*:49–73, 1970.

_____: *The Behaviour of Wolves, Dogs and Related Canids*. Jonathan Cape, London, 1971.

_____: The behaviour of cats. In Hafez, E. S. E. (Ed.): *The Behaviour of Domestic Animals*, 3rd ed. Bailliere Tindall, London, 1975.

Fuller, J. L. and Fox, M. W.: The behaviour of dogs. In Hafez, E. S. E. (Ed.): *The Behavior of Domestic Animals*, 3rd ed. Bailliere Tindall, London, 1975.

Ginsburg, B. E.: Evolution of communication patterns in animals. In Hahn, M. E., and Simmel, E. C. (Eds.): *Communicative Behavior and Evolution*, Academic Press, New York, 1976.

Harris, D. R.: Agricultural systems, ecosystems and the origins of agriculture. In Ucko, P. J. and Dimbleby, G. W. (Eds.): *The Domestication and Exploitation of Plants and Animals*. Duckworth, London, 1969.

Herre, W.: The science and history of domestic animals. In Brothwell, D. and Higgs, E. (Eds.): *Science in Archaeology*. Thames and Hudson, London, 1963.

Hilzheimer, M.: Dogs. *Antiquity,* 6:411–19, 1932.

Jolly, A.: *The Evolution of Primate Behaviour*. Macmillan, New York, 1970.

Joslin, P. W. B.: Movements and home-sites of timber wolves in Algonquin Park. *Amer Zool,* 7:279–88, 1967.

Kleiman, D. G.: Some aspects of social behaviour in the canidae. *Amer Zool,* 7:365–72, 1967.

Kretchmer, K. R. and Fox, M. W.: Effects of domestication on animal behaviour. *Vet Rec,* 96:102, 108, 1975.

Lawrence, B.: Early domestic dogs. *Z Saugetierkd,* 32:44–59, 1967.

MacDonald, D. W. and Apps, P. J.: The social behaviour of a group of semi-dependent farm cats, *Felis catus*: a progress report. *Carnivore Genet Newsletter,* 3:256–68, 1978.

Masserman, J. H.: Ed. *Animal and Human*. Grune & Stratton, New York, 1968.

Mech, L. D.: *The Wolf: the Ecology and Behaviour of an Endangered Species*. Natural History Press, New York, 1970.

Meggit, M. J.: The association between Australian aborigines and dingoes. In Leeds, J. and Vayda, P. (Eds.): *Man, Culture and Animals*. A.A.A.S., Washington, D.C., 1965.

Montagu, M. F. A.: On the origin of the domestication of the dog. *Science,* 96:2483, 1942.

_____: *Culture and the Evolution of Man*. Galaxy Books, London, 1962.

Musil, R.: Domestication of the dog already in the Magdalenian? *Anthropologie,* 8:87–88, 1970.

Olsen, S. J. and Olsen, J. W.: The Chinese wolf, ancestor of new world dogs. *Science,* 197:533–35, 1977.

Pfeiffer, J. E.: *The Emergence of Man*. Thomas Nelson & Sons, London, 1972.

Reed, C. A.: Animal domestication in the prehistoric Near East. *Science,* 130:1629–39, 1954.

Sauer, C. O.: *Agricultural Origins and Dispersals*. MIT Press, Cambridge, Mass., 1952.

Schenkel, R.: Submission: its features and functions in the wolf and dog. *Amer Zool, 7*:319–30, 1967.

Scott, J. P.: The process of primary socialisation in canine and human infants. *Monograph Soc Res Child Develop, 28*:1–49, 1963.

———: Evolution and domestication of the dog. *Evolution Biol, 2*:243–75, 1968.

Scott, J. P. and Fuller, J. L.: *Genetics and Social Behaviour of the Dog.* Univ Chicago Press, Chicago, 1965.

Smith, H. S.: Animal domestication and animal cult in dynastic Egypt. In Ucko, P. J. and Dimbleby, G. W. (Eds.): *The Domestication and Exploitation of Plants and Animals.* Duckworth, London, 1969.

Todd, N. B.: Cats and commerce. *Scientific American, 237*:100–107, 1977.

———: An ecological, behavioural genetic model for the domestication of the cat. *Carnivore, 1*:52–60, 1978.

Turnbull, P. F. and Reed, C. A.: The fauna from the terminal pleistocene of Pelegawra Cave. *Fieldana Anthropol, 63*:99, 1974.

Woodruff, G. and Premack, D.: Intentional communication in the chimpanzee: the development of deception. *Cognition, 7*:333–62, 1979.

Woolpy, J. H.: Socialisation of wolves. In Masserman, J. H. (Ed.): *Animal and Human.* Grune & Stratton, New York, 1968.

Woolpy, J. H. and Ginsberg, B. E.: Wolf socialisation: a study of temperament in a wild social species. *Amer Zool, 7*:357–64, 1967.

Zeuner, F. E.: *A History of Domesticated Animals.* Harper & Row, New York, 1963.

CHAPTER 2

RELATIONSHIPS BETWEEN THE HUMAN
AND NONHUMAN ANIMALS*

Michael Fox

What value is there to the veterinary profession in analyzing the pet–owner bond, and indeed what benefits might thereby be gained for the animal, for the client/owner, and for society as a whole? This book certainly demonstrates that there are indeed many benefits, real and potential, to be derived from objective, multidisciplinary studies of the pet–owner bond. In addition to the "whys" of such research, some methodological "hows" are detailed, and one may predict that more sophisticated techniques and analytical procedures will be developed once the significance of this field is more widely appreciated.

Some of the following general questions will be answered in this chapter and others, while other questions may be worth further investigation in the future.

1. What cultural or psychohistorical factors influence people's attitudes, values, and perceptions toward animals? (Shepard, 1978) (*See* Chapters 4 and 5 in this book.)

2. What cross-cultural, socioeconomic, and religious factors influence the pet–owner bond? (Levinson, 1972) (*See* Chapter 1 in this book.)

3. What other developmental, social, sexual, and emotional factors determine an individual's animal related attitudes and preferences? (Levinson, 1972) (*See* Chapters 5 and 10 in this book.)

---

*Portions of this paper are taken from the author's forthcoming book, *The New Eden: Animal Rights and Human Liberation*, Viking, New York. Used with permission.

23

4. Why do people keep pets and what are the motives and rewards? (Fox, 1980) (*See* Chapters 4 and 5 in this book.)

5. What are the reasons for people abandoning, mistreating or otherwise abusing their pets? (Fox, 1980) (*See* Chapter 5 in this book.)

6. How does the personality/emotionality, behavior, and life style of the person affect the pet (behaviorally, physiologically, healthwise, etc.) and vice versa, e.g. petting-evoked bradycardia? (Fox, 1978; Lynch, 1970) (*See* Chapters 3, 6, 15, and 16 in this book.)

7. What family and interpersonal dynamics affect the primary owner (caretaker's) bond with the pet and influence the pet's behavior and health? (Tanzer, 1977) (*See* Chapters 15 and 16 in this book.)

8. What influence does the presence or introduction of a new family member (child or new pet) or the departure or death of a family member have on the pet and on the pet–owner bond? (Levinson, 1972) (*See* Chapters 17 and 18 in this book.)

9. What effect does short-term separation, death or sickness (of either pet or owner) have on the survivor or relationship? (Fox, 1980) (*See* Chapters 16 and 18 in this book.)

10. What channels of communication are utilized in the pet–owner relationship and how well understood, effective, and accurate are they for interspecies communication? (Fox, 1974 a; 1974 b) (*See* Chapter 3 in this book.)

11. What are the optimal criteria for insuring that the pet–owner relationship will be satisfying and fulfilling for person and pet alike? (Fox, 1980)

HUMAN AND NONHUMAN ANIMAL RELATIONS— A HISTORICAL OVERVIEW

Archaeologists and anthropologists estimate that since the evolution of our species, we have spent nine-tenths of our past as hunter-gatherers (Lee and Devore, 1968). Only recently, within the last 10,000 years, have we domesticated plants and animals and within the last 200 years developed technologies. Our earliest relationships with and "uses" of animals were therefore very much like that of other predators, such as the wolf and lion. We had to kill them for food. Later, when we developed the skills, we utilized their remains for many purposes: skin and sinew for clothing, thread, and fishing lines; and bone for tools and weapons.

Man the Hunter

From the pieces that can be fitted together from our long past as hunters, our forefathers were not primitive savages, half hyena and half ape. Their consummate skills in living under often harsh conditions, equipped with the most rudimentary "technology" of bone, flint, and later metal instruments, attest to their intelligence and adaptability. Furthermore, their exquisite cave paintings and carvings, their often elaborate burial sites, and later astrological calendars (such as Stonehenge) indicate that they possessed a sensitivity toward nature and a wisdom of the earth and cosmos that may well surpass our own today.

Man's relationship with animals as a hunter, idyllic in many ways judging from the cultural tradition and attitudes of the few remaining hunting societies of today, eventually came to a close some ten to twelve thousand years ago. With uncanny synchrony throughout the world, possibly linked in some way with the end of the last great ice age, agriculture emerged. Plants and animals were domesticated, tamed, herded, corralled, and even selectively bred for desirable traits. Natural meadows were ploughed up for crops, forests cleared and swamps drained for raising more crops and livestock. The Garden of Eden was now being changed because man had become its gardener.

Man the Domesticator

As domesticated plant and animal species and farming techniques were traded and spread throughout the world, more and more agrarian communities were established. Game and natural habitat were less and less available in more heavily developed regions, except to wealthy land owners who could "afford" to keep some game on their estates. By the Middle Ages, much of Europe was under the plough and axe. Hunting preserves, such as the deer parks of England, were set aside for the rich aristocracy. Hunting, once a subsistence activity practiced by all, became an exclusive luxury for the upper class.

As man changed the natural world, so his relationships with animals changed. Wild animals that were of no "use," either as game or for domestication, were exterminated. The concept of wildlife management began in the game parks and hunting preserves. Carnivores, such as wolves, wildcats, and martens, were some of the first to be exterminated as "competitors." This tradition is still carried on in the United States today, where game is still relatively more plentiful than in Europe and is thus available to more than the aristocracy.

It is important at this stage to realize that when man became the gardener of Eden, he did not abandon his earlier subsistance hunting activities. The basic need to hunt persisted, adding to the escalating destruction of wild animals. Game hunting (and falconry) was elevated to the "sport of kings"; hunting per se, once a subsistence necessity, became a nonessential recreation.

During the transition from hunting to farming, humanity acquired the resources and freedom to develop not only a rich diversity of cultural activities but also a relatively new relationship with animals, namely recreation. Many of these recreational pastimes had elements of earlier hunting practices, such as coursing and bear baiting. Others, such as bull fighting and chasing a fox with a pack of dogs, became highly ritualized cultural traditions, sometimes echoing even earlier rituals of sacrificial slaughter.

When there is plenty, there is time for recreation and enjoyment, and the affluence of an agrarian way of life brought with it a new epoch of animal exploitation for spectator and participant "sports," such as the Roman games and British bear and bull baiting (Carson, 1972).

The fate of wildlife was sealed because we neither tempered the hunting urge nor realized that in changing the environment to raise our own crops and animals we would have to exterminate many wild species that competed with or preyed upon our own stock.

With the domestication of animals came a new regard for them. The first species to be domesticated is thought to be the dog, most likely derived from a southern wolf and/or a dingolike dog. Only later were other animals domesticated. The Egyptians were perhaps the first to experiment extensively with a wide variety of species, many of which proved to be of little use. While some were raised primarily for food or clothing, others, such as the ox, mule and dog, were used as beasts of burden. Later, selected species were raised and bred for specific purposes, including hunting and related sports (cock and dog fighting) and war. Horses, mules, camels, dogs, and even elephants have played their role in man's conquest over his own kind as well as over nature (Zeuner, 1963; Dembeck, 1965).

While the underlying motives in domesticating animals were basically utilitarian, other attitudes certainly influenced the relationships between man and his animal "creations." As the survival of man the hunter depended upon an intimate knowledge of the environment and of the major prey species that were killed for food, so too man the farmer had to acquire considerable knowledge about the land and animals under his dominion. It must have been realized very early on that sensitivity and empathy were an essential part of good animal husbandry. One who did not "know" his animals and care for them humanely could not be a successful horseman, shepherd, cowman, or whatever. The Egyptians, for example, not only developed the basic scientific skills of animal husbandry, they also had a deep reverence for the animals under their care, some of them even being revered as deities (Schwabe, 1978). Such animalistic and totemistic beliefs and traditions may well have originated from our earlier hunting-gathering stage where pantheism and zoomorphic religions and superstitions were linked with a deeply felt dependence upon animals for one's survival. There must also have been a sense of deep personal investment and pride in the domesticated creations of the husbandman. This endures today in the country livestock shows, sheep and gun dog trials, horse riding contests, and dog shows.

Man the Technologist/Scientist

As man became more adept in breeding and rearing domesticated animals, techniques of animal husbandry and veterinary medicine evolved into sciences, replacing animistic beliefs and the mythos of animal oriented religions and superstitions. While as a hunter, man did not have to concern himself with knowing what to feed and how to breed animals; as a farmer he had to learn such things and more: he had to learn to recognize and treat plant and animal diseases, for his livelihood depended upon it. With greater freedom from the vicissitudes of nature that were unavoidable for man the hunter came the burden of greater responsibility. The more man changed the environment and developed domestic strains of plants and animals, the more he had to know in order to survive. There was an evergrowing population to feed, together with more nonsubsistence needs to be satisfied.

It would seem, therefore, that as man became more and more involved in domesticating animals to satisfy various utilitarian needs, the act of creating such animals gave a new dimension to man's relationship with such animals. Investment, both financial and emotional, status and pride of possession, aesthetic appreciation, and the consummate rewards of companionship (with a horse or dog) and of producing highly prized offspring added both depth and diversity to the animal–man relationship.

As his consciousness changed, so his relationship with nature and animals changed. The human species was evolving from an earlier stage of simply being a part of nature to one of being separate from nature as an increasingly independent and autonomous species. Living simply by "being" became superceded by the imperative to know. But with the advent of agriculture, man had to plan for the future and understand all the qualities and needs of those plants and animals that he was endeavoring to raise and keep for himself. The more of nature he took on, the more he had to know. In this way, the human species began to evolve an objective consciousness of nature. In other words, as man's relationship with animals and nature changed, so too did his consciousness (Shepard, 1978).

Nature, or at least those parts of it that we wished to control and exploit, became part of the accumulated knowledge of the community. The industrial revolution, with its emphasis on applying knowledge through technology to dominate and exploit nature to further human progress, was perhaps the turning point in changing our relationships toward nature and attitudes toward animals. The utopian dream of a world of plenty for all excluded the rest of creation.

Contemporary Concerns

The promise of a great "marriage" between man and nature was latent in our hunting stage. It began to blossom when we became farmers and husbanded parts of nature. But now this marriage between man and nature is being destroyed, as are our relationships with each other and with our fellow animals (Fox, 1976). We are becoming competitors and adversaries in what was once a potentially fruitful marriage. We no longer "husband" the land or our livestock; instead we manage "factory" farm agri-businesses and "harvest" whales and other wildlife "resources" (Harrison, 1964; Scheffer, 1974).

Because of these contemporary cultural transformations, our relationships with nature and animals have changed significantly. Inhumane practices are now justified for the "good of humanity" or in order to insure our survival. If this were not so, then farmers would not rationalize that in order to feed so many people, some abuse of the land and of animals is inevitable, and scientists would not say that animals must suffer under various tests so that people can have safe cosmetics and other nonessential consumables. How these relationships might be improved is considered in depth in my book, *The New Eden: Animal Rights and Human Liberation.*

Against this historical background that reviews the various stages that our relationships with animals and nature have taken, one major contemporary area of concern will now be explored—the pet–owner bond.*

THE PET-OWNER BOND

There are many subtle values and needs associated with pet ownership. Pets provide companionship for those lonely people who feel alienated from or dehumanized by society. They give a

---

*I hope that in future, the term "pet" will fall from general usage and be replaced by "companion animal," which is kept not by a "master" but by a "human guardian."

sense of family and community to countless numbers of elderly retired people and also to young couples who are childless. Pets are significant nonhuman companions and child substitutes for many. They are always accepting, and with the unconditional love that they can offer, they are beneficial for a person's emotional well-being.

Providing elderly people in retirement homes with a pet has significantly eliminated frequent episodes of depression (Mugford, 1974). People need to be needed, have someone to love, a desire which many pets—cats, dogs, and parakeets—will fulfill. The loyalty and devotion of a dog can help an insecure child or adult maintain a sense of belonging. A pet can be a nonthreatening subordinate "other" to a child and as such give the child a sense of unconditional acceptance and foster responsibility and compassionate understanding.

I also believe that one of the greatest "values" of pets, the ultimate nonutilitarian "use" of an animal, is to help people regain their "animal/nature connection": to become more fully human by an interspecies relationship that can break down egocentric and humanocentric perceptions and valuations. First, I will describe four somewhat arbitrary categories of owner–pet relationships and those values and attitudes that characterize each category.

Object-Oriented Relationships

Pet ownership today may comprise a percentage of relationships that may be termed object-oriented, as when a family has one or more dogs and cats because: "What's a home without a pet; we always had them as kids and so did our parents." Like the Easter bunny and chicken syndrome, the Christmas or birthday kitten or puppy may be nothing more or less than another play object for the children. As with any inanimate toy, interest in the pet may wane with time as its novelty fades and appealing infantile traits are replaced by less endearing or positive affect-releasing behavior and temperament. Similarly, adults may keep pets for purely ornamental purposes, the animal being regarded with much concern and empathy as a piece of furniture in proportion to its monetary worth or uniqueness. Furniture and other inanimate objects may have, however, some intrinsic worth — sentimental value or an associated feeling of identity or status. Pets may serve the same function at this next level of relationship.

Exploitative, Utilitarian Relationships

The animal is used, trained, manipulated, or exploited to varying degrees, sometimes inhumanely and unethically, for the exclusive benefit of people. This category includes the use of animals in all forms of biomedical research, in military work, and in agriculture as food converters for human consumption. In the domain of petdom, possession of an animal for any utilitarian function — as a guard, bird dog, guide for the blind, for show, obedience trials, and for breeding purposes — involves this kind of relationship to varying degrees, in combination with varying degrees of need-dependence and object oriented relatedness. The policeman who loves his working dog as a companion and the breeder who regards his best dog as a status object both combine utilitarian/exploitation with need-dependence and object oriented relatedness respectively.

The relationships between owner and pet cannot be defined simply since they may contain varying degrees of one or more of the previously mentioned types of relationships and therefore entail a whole complex of different needs, motivations, values, and attitudes.

A pet that is kept primarily as a "learning experience" is in this category of utilitarian-exploitation if the learning is not integrated with empathy. As elaborated on in my recent book, *Between Animal and Man*, this essentially utilitarian view can be one of the most destructive and dehumanizing forces that distorts our perceptions and world view. To value a pet (or a person) simply on the basis of his or her utilitarian function is crass "deanimalization."

Need-Dependency Relationships

The next level of interpersonal intimacy is to regard the pet as a source of satisfaction for various needs and dependencies. This type of need-dependency relationship is perhaps the major reason why people have pets, especially dogs and cats; it is also a common underlying emotional mode in many relationships between people. The pet may fulfill many needs for adult and child alike: a companion; "confidant"; sibling partner in many games; an unconditionally affectionate and accepting emotional support; a link with nature, with natural, uninhibited, and honest emotions and responses; a refreshing break from the vacuous impersonal and often dehumanizing human transactions. The deeper the emotional need of the person, the more important the pet becomes in this kind of relationship. A pet may therefore serve as a psychotherapeutic support as alluded to earlier, as a child or sex substitute, a guard for the paranoid, an ally for the lonely and antisocial, a being to control and subjugate for the megalomaniac, or an outlet for redirected frustration and aggression. Companion animals are gaining wider recognition and respect as therapeutic tools, serving as an emotional bridge and social catalyst between an emotionally withdrawn patient and the therapist. Even in "normal" contexts, a companion animal serves a significant role as social catalyst and emotional bridge in the home, facilitating interactions between family members and between them and house guests also.

With selective breeding, enhanced by overpermissive and overindulgent rearing, the pet may become neurotically over-attached to its owner (Fox, 1974a; 1974b). This is a source of many behavioral disorders in pets and is often a result of the owner's need to indulge some living creature, in essence as a child substitute. It gives the person a sense of belonging and for some, even something to live for, especially for senior citizens who may have no social life and no emotional fulfillment outside of the relationship with their parakeet, cat, or dog.

Death of the Pet

Because of the emotional investment and dependence, a need-dependency relationship differs significantly from object oriented and utilitarian relationships and is best exemplified by the way in which the individual or family responds to the pet when it is sick or after it has died or gotten lost. In the optimal milieu, a child can come to understand something about the meaning and significance of death as well as learn to cope with personal suffering when a loved one dies. Knowing that a pet will not live as long as he or she will and experiencing the death of a pet can force the child to face the ultimate reality, and as a growth experience, parent or teacher-facilitated, it may be a significant, indirect benefit of pet ownership for children. Knowing that all things that live will die can give a clarity and significance to relationships with other living things from which an understanding of, and reverence for all life may be fostered. Seeing the pet suffer in sickness can also develop empathy and compassion and may also help children become better parents when they mature.

The Actualizing Relationship

There is a final dimension to the social values and uses of pets that is seen in what I regard as a mature, actualizing relationship between pet and owner. Here the pet is related to essentially as a respected "significant other"; its intrinsic worth being appreciated for itself instead of for reasons of status, utility, or emotional support. The person's perception and understanding of the pet shifts from one of dependency to one that is self-actualizing, less egocentric, and more transpersonal. This change in perception and understanding has been recognized and analyzed earlier in human relationships by the humanistic psychologist Abraham Maslow (1968) and is no less applicable to man–animal relationships.

It is easier for many to establish such a relationship with an animal than with a person since the latter's insecurities, ego defenses, and associated expectations and attitudes can be an additional barrier to the person's own barriers. These are absent or at least minimal in a pet (or young child), facilitating the establishment of a natural, transpersonal relationship, but more often, this unconditional openness and receptivity in the pet (or child) can lead to a need-dependency relationship, satisfying such human needs as to mother and indulge or dominate and control.

A longtime trainer of German shepherds recently told me that he had started to "ease off" (relinquishing some of his need to control others) and was amazed at what he has learned about his dogs. They were becoming his teachers, and the one way man-animal communication had shifted into a more reciprocal relationship with obviously beneficial results. Boone (1954) lucidly describes this change in man-animal relatedness.

Misguided "Naturalism"

As in caring for children, the adult in relating to an animal has a degree of care and responsibility that must not be surrendered in a permissive ideal of natural laissez-faire, e.g. the misguided "naturalism" of letting dogs be dogs and allowing them to run free, breed indiscriminately, etc. With increasing maturity of the child and to some degree of the pet, responsibility ideally shifts from the parent/owner to the child/pet.

One cannot expect a pet to mature to this degree of responsibility in relation to others (obedience training helps to some extent), and it is important for those people who have this kind of quasi-naturalistic relationship with their pet not to lose sight of the social and ecological aspects of responsible pet ownership (*See* Chapter 13). Living easily with a pack of dogs in the hills may be a nice Thoreaulike pastoral idyll, but such animals may be shot by hunters, fall into traps, eat poisoned bait, compete with natural predators for prey and carrion, kill domestic livestock, contribute to automobile accidents, and even attack people. Just as with a child, the pet must be protected. A natural relationship without responsible ownership is to be deplored. Some hold that a pet should be allowed to live as natural a life as possible, to reproduce freely, and to roam and hunt as it pleases, be it a dog or a cat. Such ideals are contrary to the contemporary socioecological ramifications of pet ownership, and while some may need to be educated toward greater empathy with their pets, others need to be educated toward greater responsibility.

Responsible Stewardship

One final and almost terrifying relationship with animals and nature as a whole is emerging today, entailing an almost Godlike awareness and responsibility: *stewardship*. This is to be distinguished from "management," which, even under the best guise is nothing more than utilitarian exploitation. Stewardship entails constant monitoring of every human interaction in any biological system including natural ecosystems, ranging from such things as oceans to wilderness habitats and man-made systems from the soil of our farm lands, the genetics and metabolism of high yield dairy cattle, to the gait and temperament of a German shepherd. The latter is a good example of how natural selection pressures have been relaxed and replaced by selection criteria, some of which may be detrimental to companion animals. Breed clubs of pure-bred varieties of cats and dogs are now beginning to identify, monitor, and correct some of the destabilizing consequences of domestication and unnatural selection. The physical and psychological degeneration and emotional problems of our companion animals necessitate our most concerted stewardship. Also, since some of the behavioral problems that companion animals develop are linked with emotional troubles within the family or stem directly from the owner, family therapy involving the animal as well is now becoming a reality.

SOME CONTEMPORARY PROBLEMS

Contemporary problems other than the issues of animal over-population and irresponsible ownership threaten to change long-existing relationships between man and his animal companions. Many high-rise apartments and condominium units outlaw any kind of animal, a situation analogous to racial zoning and other discriminatory attitudes and laws governed by consensus. Surgical modifications such as declawing, neutering, and debarking are often indicated today in order to preserve the pet–owner relationship and to appease or otherwise satisfy social pressures. The ethics of such practices are a cause of considerable concern. A noisy and untrainable dog that has been devocalized might, in the final analysis, be better off than dead, though some would contest this view. The question remains, however, as to just how far we should go in making our companion animals adapt to our needs and life-styles.

No companion animal should be allowed to roam free today, not only for its own safety but for the many social and ecological reasons described earlier. How they can adapt to a more confined life-style is a moot question. Selectively breeding to make them even more adaptable (and perhaps even more unnatural) is another serious ethical issue. Some would hold that the cat is the animal of the future since the dog (other than a toy breed) is no animal for the urban environment. Few owners will pick up after their dogs have defecated, and the time has come when dogs are no longer allowed to run free in city parks and are not even allowed into some parks whether on or off the leash.

Although many dogs may be urban "misfits" by temperament, the changed life-style of the family can be an aggravating factor; few dogs adapt well to being left alone in an empty house where both husband and wife go out to work. They can become house wreckers, bark excessively, or become unhousebroken. Dog "walking" and "sitting" and dog day-care centers have been established and provide useful service here.

Ethical and Humane Concerns

Keeping a companion animal to fulfill some emotional need for companionship, security, protection, status, etc. is ethically acceptable provided that in the relationship the pet does not suffer physically or psychologically, such as being fed an improper diet or indulged to become overdependent or undisciplined and unstable and potentially dangerous to other people or children. The relationship may also be ethically unacceptable if the owner indulges certain of his or her own fantasies or needs on the animal, such as excessive control or punishment, using the animal as a scapegoat or focus for redirected aggression, or as a sexual outlet. It is also unacceptable when owners choose to project or seek an outlet for their own repressed needs onto the animal, believing that it should live a "natural" life — free to roam uncontrolled and free to breed indiscriminately. Emotional factors do underlie many cases of irresponsible ownership as distinct from perhaps more widespread indifference and ignorance. An owner may identify his or her own sexuality in the companion animal and see neutering of it as a personal threat. The relationship based upon emotional need is also unsatisfactory if it impinges upon the rights of others (such as having a noisy dog or an unrestrained pet) or if the animal is a threat socially or ecologically.

The ethical question of how far we should go to make an animal adapt better to certain human contexts, such as apartment living, remains to be resolved. It is not always a purely personal decision whether or not to neuter, declaw, or devocalize a dog or cat since social and ecological factors are also involved. Provided no other alternatives are available or are effective (training, genetic selection, etc.) and the alteration to the animal is done humanely and there is no postalteration suffering physically or psychologically, then it may be an ethically acceptable decision, certainly preferable to relinquishing or destroying a pet.

Responsible compassion and rational humane killing are often necessary with animals today, and emotions tend to get in the way of this necessary stewardship. The humane destruction of millions of unwanted dogs and cats, the killing or sterilizing of physically and emotionally unsound purebreds, are responsible actions often abdicated by profit motivated breeders and sentimentally misguided humanitarians. Culling (as in natural selection where "Mother Nature" maintains the quality of life) is a responsibility man cannot abdicate in propagating domestic animals if they are to be sound both physically and psychologically. But having genetically sound animals is only the first step. Owners should also be educated as to the proper care, rearing, socialization, and training of their companion animals and also be aware of the social responsibilities of owning an animal.

Animal Expectations

People often have illusory and unreal *expectations* about companion animals, knowing and caring little about their basic needs and natural behavioral tendencies. Dogs like to bark and to roam. They don't like to be left alone in an apartment for extended periods. Cats like to claw furniture, sometimes to spray the house, and to go out a courtin' and a caterwaulin'. Human expectations can get in the way of appreciating the animal for itself, independent of the owner's needs and demands. Keeping a companion animal can help one mature through understanding, to appreciate the intrinsic worth and basic rights of a fellow earth being. As we learn to relate better with animals, so will we relate more effectively with our own kind since selfish demands and unrealistic expectations are a barrier to any form of meaningful relationship, be it with a human or with a nonhuman being.

Pet owners are frequently chastized for treating their animals like little people, buying them nonessential accoutrements and being wholly anthropomorphic in their perceptions of their pet (Szasz, 1968), believing that their pet feels and thinks just like any human being would. I find nothing wrong with the former, provided the indulged animal does not suffer physically (by becoming neurotically overdependent). Many animals bred and raised for this role thrive on it, and for many owners they provide much needed emotional satisfaction and are of inestimable mental health value. It should not be forgotten that some people may become so pet involved that there is a pathological overattachment and introversion. Satisfied by the pet, the person ceases to make the effort to go out and seek satisfying relationships with other people. Companion animals may become the "emotional slaves" of those people who need an animal to dominate, control, overindulge, or serve as a person substitute. An owner's sense of inferiority or insecurity may be relieved by the possession of a show dog or trained attack dog.

I do not believe there is much wrong in an owner being somewhat anthropomorphic in his or her regard for the pet since there is now good scientific evidence to support the contention of many owners that cats and dogs have emotions and sensations comparable to our own—fear, pain, anxiety, jealousy, guilt, joy, depression, anger. The brain centers mediating such states are virtually identical in man and other mammals (Fox, 1974a; 1974b). This is why for educational reasons in the most general, yet most total and perfect sense, I support people keeping animals as "pets," as companions, cotherapists, and indeed, as teachers in the real sense. Companion "pet" animals can make us more human, more compassionate, patient, understanding, and responsible humane stewards. They can take us out of unhealthy humanocentric preoccupations and away from false priorities and pseudopriorities.

They, like the very earth, can keep us grounded in the one reality that is life. Through this "animal connection" we may discover our own animality and at the same time fulfill our godlike obligations to be responsible, compassionate stewards of animals. We have yet to become humane stewards, cocreators, working in harmony with the rest of nature. The mistakes that have been made with our companion animals and animals in general — the abuse, excesses, neglects, exploitation, and genetic deformities and emotional problems — are all lessons at their expense. They are our mirrors, and as we learn from them and from our mistakes so may we gain insight and grow. Many animal owners are already "there" with their pets and with the world, but countless others have yet to mature. When we all give equal consideration to animals' rights, (Morris and Fox, 1978) we will indeed be more human and civilized. Without companion animals in our lives, and especially during the critical years of childhood, our lives would surely be impoverished.

REFERENCES

Boone, J. A.: *Kinship With All Life*. Harper and Row, New York, 1954.

Carson, G.: *Men, Beasts, and Gods*. Scribner's, New York, 1972.

Dembeck, H.: *Animals and Men*. The Natural History Press, Garden City, New York, 1965.

Fox, M. W.: *Understanding Your Cat*. Blond & Briggs, London, 1974a.

Fox, M. W.: *Understanding Your Dog*. Blond & Briggs, London, 1974b.

Fox, M. W.: *Between Animal and Man*. Coward, McCann, New York, 1976.

Fox, M. W.: *The Dog: its Domestication and Behavior*. Garland Press, New York, 1978.

Fox, M. W.: *How to be Your Pet's Best Friend*. Coward, McCann, New York, 1980.

Fox, M. W. and Morris, R. K.: *On the Fifth Day*. Acropolis Books, Washington, D.C., 1978.

Harrison, R.: *Animal Machines*. Vincent Stuart, London, 1964.

Lee, R. and DeVore, I.: *Man the Hunter*. Aldine, Chicago, 1968.

Levinson, B.: *Pets and Human Development*. Thomas, Springfield, 1972.

Lynch, J. J.: Psychophysiology and development of social attachment. *J Nerv and Mental Diseases, 151:* 231–44, 1970.

Maslow, A. H.: *Toward a Psychology of Being*. Van Nostrand, New York, 1968.

Mugford, R. A. and M'Comisky, J. G.: Some recent work on the psychotherapeutic value of cage birds with old people. In Anderson, R. S. (Ed.): *Pet Animals and Society*. Bailliere Tindall, London, pp. 54–65, 1975.

Scheffer, V. B.: *A Voice for Wildlife*. Scribner's, New York, 1974.

Schwabe, C. W.: *Cattle, Priests and Progress in Medicine*. University of Minnesota Press, Minneapolis, 1978.

Shepard, P.: *Thinking Animals*. Viking Press, New York, 1978.

Szasz, K.: *Petishism. Pets and Their People in the Western World*. Holt, Rinehart and Winston, New York, 1968.

Tanzer, H.: *Your Pet Isn't Sick*. E. P. Dutton, New York, 1977.

Zeuner, F. D.: *A History of Domesticated Animals*. Harper and Row, New York, 1963.

CHAPTER 3

INTERACTIONS BETWEEN PEOPLE AND THEIR PETS:
FORM AND FUNCTION

AARON HONORI KATCHER*

Despite the great antiquity of pets (Davis et al., 1978), despite their enormous number in Western Europe and the United States (Schneider, 1975), despite the billions spent yearly on their care, there have been almost no studies applying current experimental, anthropological, ethological, or sociological methods to the study of the phenomenon. The published literature consists mostly of case reports and clinical incidents, which for the narrators illustrate, in some fashion, the intriguing and salient nature of the transactions between people and their pets. The work that has generated the most public excitement has been the pioneering clinical experiments of Levinson (1969, 1972), Corson et al. (1975), and Condoret (1973, 1978), in which a companionable animal was used to alter, in some way, the life of an emotionally disturbed child or adult. Similar brave and tender efforts to find a bridge to damaged or isolated children have been made by the groups and institutions that teach horseback riding to the mentally compromised or physically handicapped (Curtis, 1979). These interventions may demonstrate the inherent attractiveness of pets, but they do not tell us very much about the dynamics of the bonds that tie the very large mass of people who own pets to their pets.

I think there are very good reasons why man's pets have been pushed into such a small corner of our scientific consciousness. It is possible to identify a number of intertwined cultural themes, some of great antiquity, that have contributed to this selective

*Supported in part by USPH Training Grant: "Veterinary Medicine and the Study of Human Behavior" 5-T24-MH15400-02.

41

inattention. These cultural defences against taking a serious look at the meaning of pets are worth describing because they *still* present barriers to research and the support of research.

DRAWBACKS OF BIOLOGICAL TRADITION

For most of man's cultural history, his behavior was assumed to be regulated by moral rather than psychological imperatives. Within the Judeo-Christian moral universe, man's behavior was understood in terms of his relationship to other men and to God. The natural environment, including both its animate and inanimate components, was at best morally neutral and more usually malign. Out of this anthropocentric moral tradition we have built a psychological tradition in which all significant events in a person's life are seen as transactions between people. Moreover, the set of psychologically significant people is usually rather small, frequently restricted to the members of the nuclear family.

Our tradition of "scientific psychology" is no less anthropocentric than our moral tradition but in fact is even more limited than our moral heritage in that it has excluded both the natural and the supernatural from its vision.

In part, the failure of our scientific view of man to widen the anthropomorphic bias of our moral tradition is a result of the metaphor that René Descartes used to fashion a morally neutral universe into which the seventeenth century could expand (Descartes, 1960). By describing animals as machines, a description that also applied to the animal part of man, he permitted both human beings and animals to be abstracted out of any natural environment. Machines operate anywhere provided that the interior requirements for their operation are present. Since feelings are part of the machinery, a psychology of feelings can be construed by studying man using no other context save an internal one.

This Cartesian nullification of the environment was incorporated into the tradition of nineteenth century experimental medicine. The clinics of eighteenth and nineteenth century France, which were the maternity hospitals for the modern medical sciences, saw their patients isolated from home and community (Foucault, 1973). The experimental physiology of Claude Bernard isolated the animal by anaesthesia to remove the influence of the environment on the central nervous system and permit the study of internally regulated events. It was the need for the isolation of the subject as much as human concerns that mandated the use of anaesthesia in experimental physiology. The great psychologists of the late nineteenth and early twentieth century—Freud, Pavlov, and Cannon—were all nurtured in experimental physiology. All of them removed their subjects from the environment.

Initially, Freud used hypnosis to isolate his subject from all stimuli except internal ones. Later, the position of the patient on the couch served the same purpose. Pavlov used a chamber to insure that the dog attended only to those stimuli that were presented by the experimentor. Cannon's work is significant to us largely because he developed *internal* biological criteria for the identification of emotional states in isolated animals (Cannon, 1915). These criteria—changes in heart rate, blood pressure, blood sugar, and catecholamines—were then used as criteria for emotional expression in human beings also studied in isolation. To this day almost all of our current psychological research in animals and people is still a study of individuals isolated within some kind of artificial framework.

HOW TO PET

Such a Cartesian tradition looks upon interactions, like the interactions between people and pets, with a worried eye. If you have two persons interacting it is not possible to control the behavior of one precisely so that one can be a "stimulus" for the other. Most so-called "social psychological" experimentation is really the description of the effect of trained stooges upon a single subject. I can best illustrate the concern about controlling stimuli that dominates classic experimental psychology by describing the comments of James Lynch, a colleague in these researches, from our graduate psychology seminar. He was presenting his data about cardiovascular changes that occur in dogs when they were patted by human beings (Lynch, 1966). It was the unanimous conclusion of the graduates that he should build a petting machine, a mechanical instrument to stroke the dog. With such an instrument it would be possible to precisely "control" the event. These psychologists could not comprehend that the machine would of course eliminate the phenomenon being studied, the dialogue between man and dog.

The biological tradition of noninterventive observation would seem to stand in opposition to the analyses of the experimentalists. The work that placed human and animal emotions in a biological context, Charles Darwin's *Expression of Emotions in Animals and Man* (1913), provides elegant descriptions of interactions between people and their pets. Unfortunately, the philosophical underpinnings of evolutionary biology tended to force people's attention away from such mundane events as a dog enjoying the caress of its master. Evolutionary biology was closely wed to nineteenth century utilitarianism, and social Darwinism was as important an intellectual movement as biological Darwinism. Evolution depended on competition for survival, and survival depended upon the necessities of life. The animal psychology that grew out of this context focused upon the attractions of food and the fear of pain. Appetite and fear became the pillars of experimental and biological observation.

It has been only within the last twenty years that we have been able to enlarge our concept of the psychology of infancy to include more than orality and fear. One of the most exciting developments has been the discovery of affection and the importance of touch and the comfort provided by contact with other beings. Spitz (1946), Bowlby (1973), and Harlow (1959) left us with a new economy of infancy. The discovery of the importance of affection in adults has come slowly, again because of the nineteenth century tendency to subordinate affection to sexuality. Sexuality has the virtue of being a prize in evolutionary warfare and, within the morality of the time, a scarce economic good. We are still trying to look at affection between adults with a new eye and hesitantly trying to examine touch as something more than foreplay.

EVOLUTIONARY BIAS IN ETHOLOGY

Modern ethological methodology, which would be an ideal tool with which to study people and their pets, owes much to

Darwin's study of emotions. Yet the evolutionary context within which Darwin framed his observations made it difficult for the first field of ethology to focus on mankind's interaction with animals. The difficulty was provided by the emphasis within the field of ethology of observations made under natural conditions. It is not possible here to trace the vicissitudes of the concept "natural" in our scientific thinking. It is only necessary to note that the justification for the onerous study of animals in the wild was the recognition that animals confined in experimental settings or in zoos behave very differently from animals observed in their own habitat.

The "zoophobic" bias of ethology was reinforced by the recognition that any cultural context suspends the inexorable play of evolutionary combat. A natural context is one in which animals are competing for survival against other animals. People's pets are domesticated animals living in a context far removed from anything that could be called a state of nature. They are protected and forbidden to compete for survival. Obviously pets, the most insulated and unnatural of domestic animals, are only marginally interesting as objects of observation. The tendency of amateurs of ethology to idolize the wolf and denigrate the dog as an offshoot of a once noble animal is an example of the power of our feelings about the superiority of beings still "in nature."

Anthropology has provided excellent studies of the symbolic importance of animals in unindustrialised societies. Yet there are almost no studies of the cultural roles of animals in urban societies. The conceptual tools of anthropology seem to operate better when the anthropologist spends his time with the afflicted either within our own societies or within the people of the third world. Ironically one could say that the anthropologists respect the rules of comportment in capitalist society. They do not intrude upon the privacy of the rich or even the moderately well to do. Their attentions are almost exclusively confined to the poor, powerless, or afflicted.

Anthropology, in its focus on the primitive and afflicted, mirrors the gaze of psychology in its focus on pain, trauma and the primitive and unacceptable urges that originate from the unconscious. There is always something that reminds one of soap opera and gothic novels in the case reports of psychologists and analysts. Like the television audience, they dote on violence, sexuality, and affliction: catastrophe of one sort or another. Our life with animals fits much more into the category of the sentimental rather than the primitive or the violent, and we have no psychology of the happy and the mundane. There the blindness of science is combined with the romantic bias still inherent in our definition of good taste in art and literature.

We are all taught to despise the sentimental, to think of it as banal or as a cover for darker hidden emotions. Thus, the sentimentality of German culture piqued our interest because it was felt to mask the rage that was unleashed by the Third Reich. Affection for puppies is of interest only when the person demonstrating the affection is a concentration camp guard or an ax murderer (Morris, 1973). Sentimentality in art or the novel implies a trivialization of emotion, a blindness to reality, or an absorbtion into bourgeois material living. True art is as much characterised by pain and passion as is true science.

SENTIMENTALITY

Unfortunately, it is also probably true that people's response to pets has the character of sentimentality. Puppies bring forth a stock emotional response that, however strong, does not alter the life of the observer and does not afflict or change him in any way. It is a stock response, common from one person to the next, and always or almost always saccharine and unalloyed with any darker feeling. If one reads the stories about pets that find their way into newspapers, children's books, or magazines with a large circulation, their tone is almost always sentimental. Scientific accounts of the animals helping people have the same feeling tone. Accounts of "seeing heart dogs," the writings of Corson, Levinson, and Condoret, all fall into this category.

Lest you think that I am trying to denigrate this style of description, let me quote a letter of Freud describing the relationship between dogs and people (Freud, 1976).

> "It really explains why one can love an animal like Topsy (or Jo-fi) with such an extraordinary intensity: affection without ambivalence,

the simplicity of a life free from the almost unbearable conflicts of civilisation, the beauty of an existence, complete in itself. And yet, despite all divergence in the organic development, that feeling of an intimate affinity, of an undisputed solidarity. Often, when stroking Jo-fi, I have caught myself humming a melody which, unmusical as I am, I can't help recognising as the aria from *Don Giovanni*: "A bond of friendship unites us both . . ."

Quite the contrary, I think that through a study of animals we can bring the sentimental back into focal vision and recognise its importance. We have hidden from our scrutiny those objects which evoke benign positive feeling almost in the same way that we once repressed sexual feeling.

I hope we can begin to understand the role of pets as well as that of other objects of sentimental value: plants, gardens, sunsets, snapshots, souvenirs and all those accessories to life that we currently denigrate as trivial.

I have dwelt at length upon a nexus of cultural themes that still serve to blind our perception of the role pets play in our lives. I did so because I am professionally committed to a belief that insight helps us to avoid repeating past mistakes and because I have encountered a good deal of resistance to undertaking serious consideration of pets, resistance that has been difficult to dissipate. Part of the resistance has been internal. It has been over ten years since I have known that petting a dog produces dramatic cardiovascular responses in the animal. Yet it was only this past summer that I started measuring heart rate and blood pressure in the people doing the petting.

BONDING AND HEART DISEASE

Having described some sources of resistance to research into the bond between people and animals, I would like to share one strategy for overcoming the resistance and studying the phenomenon. The strategy had its inception in a research design used to study the impact of bonds *between human beings* on disease. Recognizing that people who were married had lower death rates than people who were single, widowed, or divorced (Lynch, 1977), we wanted to study the effect of social isolation on the mortality of patients who had been hospitalized for coronary artery disease.

We had an inventory of fifty or so items relating to social support and isolation that constituted part of an initial interview. We included a question about pets as one of those items for three reasons. (1) We were interested because of previous work with touch in nonverbal sources and affection, (2) people describe themselves as feeling less lonely when they have a pet, and (3) pets are considered as a social asset. In previous research in the ethology of virus disease we had developed an instrument called the social assets scale (Katcher, et al., 1973; Luborsky, 1973). The instrument was based on the hypothesis that people who are wealthy and wise are also healthy and that possessing any attribute that the society values is likely to make you less vulnerable to disease. Pets are certainly part of the good life and deserve to be recognized as part of an inventory of social assets.

At the inception of that study our major interest was the impact of human relationship, and we expected pets only to add a very small increment in predictability to our other variable— perhaps explain the ability of some people who were isolated from human contact to enjoy continued good health.

The results of that investigation have been reported elsewhere (Friedmann et al., 1978, 1979b), and I only wish to summarize them here as a basis for the continuing discussion.

In our white subjects the presence of a pet was the strongest social predictor of survival for one year after hospitalization.

TABLE 3-I

RELATIONSHIP OF PET OWNERSHIP TO ONE YEAR SURVIVAL

Patients	No Pet	Pet	Total
Alive	28	50	78
Not Alive	11	3	14
Total	39	53	92

Chi Square = 8.9 p ‹ 0.02

Since we had so much social data and since we also had an index of physiological disability, it was possible to examine this crude finding with multivariate analysis.

TABLE 3-II

DISCRIMINANT ANALYSIS RESULTS

Variables	Variance Explained	Significance of Addition*
	%	(p ‹)
Physiological Severity	21	.00009
Physiological Severity + Pet *Ownership*	23.5	.004
Physiological Severity + Pet + *Age*	24.9	NS[†]
Physiological Severity + *Age*	21.9	.014

*Significance of addition is for the addition of the underlined variable
[†]NS = not significant at p ‹ .05

In brief, the effect of pets was not due to the better health of the pet owners at the start of the study. It was not due to better social status of the pet owners. Pet owners did not have significantly more social assets than people without pets. It was not due to the presence of dogs alone, dogs that might have had an effect on health through the facilitation of exercise. People with pets other than dogs enjoyed a better rate of survival. Most importantly *the effect of pets was not present only in those people who were socially isolated; it was independent of marital status and access to social support*

TABLE 3-III

RELATIONSHIP OF OWNERSHIP OF PETS OTHER THAN DOGS
TO ONE YEAR SURVIVAL

Patients	No Pet	Pet	Total
Alive	28	10	38
Not alive	11	0	11
Total	39	10	49

Fisher's exact test p ‹ 0.05

from human beings.

This research finding suggests to us that pets may have important effects on the lives of adults *that are independent of and supplementary to human contact.* Perhaps pets are not substitutes for human contact but offer a kind of relationship that other human beings do not provide. The research also suggests that we should look at the health effects of pets on the mass of people who have no stigmatizing mental or social disability: that is, people who are defined as enjoying good mental health and who are, to some degree at least, socially integrated. It also suggested that the study of the effect of pets on physical rather than mental disease might be rewarding. Pursuant to these conclusions we reviewed the social functions of pets that would be expected to increase longevity and decrease the incidence of degenerative disease.

COMPANION ANIMALS AND HUMAN PHYSICAL HEALTH

We identified seven functions of companion animals that would be expected to have some influence on physical health. (Katcher et al., 1980).

The first three of these would be expected to decrease depres-

TABLE 3-IV

FUNCTIONS OF COMPANION ANIMALS THAT WOULD BE EXPECTED
TO HAVE AN INFLUENCE ON THE HEALTH OF ADULT POPULATIONS

1. Something to decrease loneliness

2. Something to care for

3. Something to keep you busy

4. Something to touch and fondle

5. Something to watch

6. Something that makes you feel safe

7. Something to provide a stimulus for exercise

sion and feelings of loneliness and social isolation. The second three would be expected to decrease anxiety and automatic arousal. Any factor that decreases or prevents feelings of depression, anxiety, loneliness, and helplessness would be likely to have a positive effect on health and decrease the incidence of a broad spectrum of chronic diseases including, most importantly, degenerative cardiovascular disease. The epidemioloical findings that would lead one to expect that companion animals have the ability to increase health in this fashion have been reviewed in another paper (Katcher et al., 1980).

I would like to emphasize the strategic value of measures of physical health as a means of testing the importance of the companion animal bond among other social relationships. I do not, however, wish to suggest that dogs are the latest variety of miracle drug, following upon Vitamin E, Transcendental Meditation, Transactional Analysis, high fiber diet, acupuncture, and jogging. At most, we would expect the presence of a companion animal to explain a very modest part of the variance in incidence of integration and so forth. What I suggest is that epidemiological measures have an advantage for testing the strength and meaning of relationships with companion animals.

It is sometimes difficult to determine if a person discharged from a mental hospital is any better than when he entered. It is also obvious that people entering mental hospitals were very different from the general population before they entered. People who will develop cardiovascular disease make up over half the population, perhaps as much as three-quarters of the adult population, and hence are much more likely to resemble the general population than the mentally ill. Moreover, it is always possible to determine with reasonable accuracy whether a person is living or dead and with somewhat less certainty whether a subject's blood pressure has risen or fallen, if he has suffered a second infarct, or has increased or decreased his exercise tolerance. It is also inexpensive to add one or two questions about pet ownership to continuing epidemiological studies, and potential return in information is quite high. The most rapid and economical way to advance our understanding of the companion animal bond would be to institute such symbiotic research. I should add also that any research into the social origins of illness is seriously faulted if it does not take into account the presence or absence of pets.

<h2>SPEECH</h2>

Although it is important to demonstrate the strength of the companion animal bond by epidemiological studies of health, it is even more interesting to study the transactions that structure that bond. That kind of structural analysis should prove to be most valuable for our understanding of human communication and the function of touch, play, and certain styles of speech in human life. One way to begin such a study is to start with the origin of speech. In the beginning man talked to the animals and the animals talked to him. Almost all myths of the Golden Age have both man and animals sharing the pleasures of conversation. The stories with which our children begin to structure their imagination always contain animals that talk to human beings. With television, children observe this conversation directly, and even without televisions many children are urged to talk to their teddy bears or their pets as if the animals understood them.

What should not be surprising then is that most people talk to their animals in adult life and believe that their animals understand them. In a study of clients of our veterinary school conducted by Ms. Melissa Goodman (Katcher et al., 1980), we found that over 98 percent of the subjects said that they spent time talking to their animals, and over 80 percent said that they talked to their pet "as a person" not "as an animal." Even more surprising, 28 percent said that they confided in their animal and talked about events of the day. Do the animals understand?

Over 70 percent felt that their animal was sensitive to human moods and feelings. We did not ask at that time if the animal understood words and sentences or how many words they understood. The oversight was unfortunate. Since that time, in interviews with dog owners, I have again been surprised to find that some people believe that dogs have the capacity to recognize a large number of human words. One subject, a professor of laboratory chemistry at my university, said that her dachshund had a recognition vocabulary of 5000 words. She was both sane and

TABLE 3-V

HOW PEOPLE TALK TO THEIR PETS

People talking to pets	99%
People talking to pets as they would to a person	94%
People confiding in pets	28%
People talking about events of the day to pets	28%
People thinking pets are sensitive to their feelings	81%
(Number of subjects = 120)	

serious. Given then, that people talk to animals and believe they are in part at least understood, given that they may spend as much as four hours a day in contact and conversation with their animal (Adell-Bath et al., 1978), or that they may feel that their animal is their only confidant or only social contact (Adell-Bath et al., 1978;) (see chapter 10), what form does that conversation take and why do people find it so desirable to talk to "dumb animals?"

Motherese

One way to answer the question is to look at the form of the language used by people to talk to pets. Two linguists at the University of Pennsylvania are examining the grammatical form of such talk (Hirsh-Pasek et al., 1980). The reason for their investigation can illustrate the potential importance of investigations of the human–companion animal bond. They are testing a hypothesis about a kind of talk known as "motherese." "Motherese" is the language used by mothers to talk to babies and young children, "babytalk." One standing hypothesis about "motherese" states that it is a simplified form of language used to help young children learn to speak. Now, if the motherese used by people to their dogs has the same form as the motherese used with children, then it must follow that the sole purpose behind the use of that form of speech cannot be the facilitation of language apprenticeship. It must have other functions for the speaker. The results of that project are still being analyzed, and there remains much that we do not know about the form of such a conversation. In our laboratory, however, we have been concerned about the effect of such talk on the speaker rather than its form.

Speech and Blood Pressure

Our interest grew out of a project studying blood pressure fluctuations during a psychotherapeutic hour. We asked the question "were the elevations in blood pressure and heart rate seen when patients spoke related to the content of the utterance, or to the act of speaking under those circumstances?" We set up an experimental situation in which the subject was asked to read aloud from a dull text with the experimenter out of the room, again when the experimenter was in the room, and then the subject and experimenter talking to each other. This experimentation was made possible by the use of an automated external blood pressure monitor (Lynch et al., 1980). The results were striking. Each time the subject read aloud or spoke, the blood pressure rose. Each time the subject rested silently or was spoken to, the blood pressure and heart rate fell. The content of the conversation did not significantly affect the size of the blood pressure elevation.

We were interested in this phenomenon of the physiological stress of talk, not because of psychotherapy but out of an attempt to understand more common human relationships. One of the more frequent complaints in marriages is that one or the other partner "does not talk," "won't talk things out." "There is no communication between them." The same complaints are made by parents about their adolescent children who refuse to talk to them. Part of the answer to what makes talk stressful in a human relationship can be found by looking at the attributes of the experimental situation in which we recorded blood pressure observations. All experimental subjects believe, correctly, that they are being evaluated or judged in some way. The stress of an experimental situation is a fear of such evaluation, what one investigator called "evaluation apprehension" (Rosenberg, 1965). Outside of an experimental situation or a psychotherapeutic one there are large sets of circumstances in which talking exposes you to the risk of evaluation, correction, punishment, inattention, contradiction, and unwanted advice and instruction. There is at least one kind of conversation that does not expose you to such risks, talking to animals.

We looked at people talking to animals in an inadvertant fashion. Our primary objective was to study the physiological

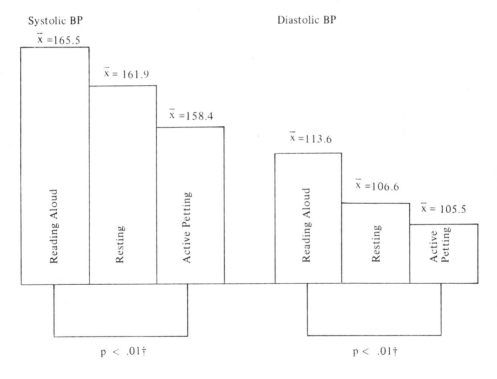

BLOOD PRESSURE*

READING ALOUD, RESTING, AND GREETING A PET DOG

Systolic BP

$\bar{x} = 165.5$

$\bar{x} = 161.9$

$\bar{x} = 158.4$

Reading Aloud

Resting

Active Petting

$p < .01\dagger$

Diastolic BP

$\bar{x} = 113.6$

$\bar{x} = 106.6$

$\bar{x} = 105.5$

Reading Aloud

Resting

Active Petting

$p < .01\dagger$

* BP recorded on person's calf with automatic Dinamap digital BP monitor

† Nonparametric Rank Difference Test

events that occur as people stroked and fondled their pets. We wanted to study the "effect of animal" on people to complement our knowledge on "the effect of person" on animals. The "effect of person" is the pronounced lowering of blood pressure and heart rate produced in an animal when it is stroked or petted. To study this phenomenon we isolated owners from their animals and measured blood pressure and heart rate using the automated blood pressure monitor. The subject was observed at rest and when he or she was reading from a section of uninteresting text. We then brought in the dog and asked the owner to pet the animal. Immediately the problems of studying a dialogue manifested themselves. The subjects did not just pat the animal, they vigorously greeted him with pats, strokes, and words. We could not, in that situation, study the effect of quiet stroking — that behavior was inappropriate to the social situation we had inadvertently created. Nevertheless, the results were revealing. During the talkative greeting, blood pressure and heart rate were significantly lower than during the time in which the subject was reading aloud. This experiment is of course a crude beginning, but the results are an invitation to continue to study the dialogue between man and animal.

Animal Therapists

We should also continue to raise the hypothesis to which this result gives partial support "Why should talking to an animal be more relaxing than talking to another human being?" We have suggested a reason. Animals do not evaluate what we say. There is at least one other. As Michael Fox suggests (1975), animals are perceived as empathic whether or not they actually are. If you remember, over 80 percent of our subjects thought their pets were sensitive to their feelings. To what degree animals are empathic, to what degree their behavior changes with the mood of their owner, is an important research question in its own right. But beginning with the recognition that they are perceived as empathic, one is immediately struck by the resemblance between dialogue with a pet and dialogue with a Rogerian therapist.

I do not mean to imply that you can not tell a Rogerian therapist from a Labrador retriever; I would like to suggest instead that the effect of talking to a dog and talking to a therapist might, under certain circumstances, be the same. One must realize of course that the animal has certain advantages. It is not possible for the client to stroke or pet his therapist and the therapist can not lick his client's hand or nose. Although I am phrasing this question with some irony, I am quite serious. One of the major contributions of Carl Rogers was the identification of empathy and a nondirective strategy as two of the essentials of a good therapeutic relationship. It is the noninterventiveness and empathy of our pets that perhaps make them such a good audience for our words.

The point that I am making rhetorically has been made empirically by the important research of Levinson, Corson, and Condoret. However, I wish to suggest that we would be limiting our focus too much if we were to think of the dog as facilitating contact only with emotionally disturbed human beings. The response of a patient to a dog or a dove in a psychotherapeutic context is only a special case of the response of people to animals in the full range of contexts in which people talk to and caress their pets. We need to explain the general phenomenon: why people find it so important and so satisfying to talk to a pet who does not attend the words being spoken.

Whatever the eventual explanation for the joy people take in talking to animals, that kind of conversation indicates the need for the serious study of a group of important dialogues in which words are used by only one participant. Within this set of dialogues I would include talking to animals, talking to teddy bears and other comforting objects, talking to the dead, and prayer. Without being irreverant, I think that it is possible to think about the similarities of the comforts of prayer and comforts of talking to an animal. Prayer is frequently accompanied by sensual enrichment, incense, music, special body postures, the touch of folded hands or rosary beads, just as dialogue with an animal is accompanied by the enrichment of touch, warmth, and odor. In both instances the talk is felt to be "understood." The analogy breaks down, of course, when one recognizes that people frequently fear the evaluation of the being to whom they pray. The resemblances between different types of one way, in the sense of words going one way, conversation are great enough that we should consider their similarities and begin to study that set of events seriously. Perhaps dialogue with animals is the most easily accessible to study.

TOUCH

Words are perhaps the less essential medium of communication between people and animals. The word that gives the name to a companion animal in English is "pet." It somehow suggests that touch is central to the meaning of a cat or dog in our lives. In previous years we had studied touch between human beings. We worked in a coronary care unit or a shock trauma unit studying the simplest kind of interaction—a nurse holding a patient's hand to take a pulse. This act could be studied because all patients on such units are monitored electronically. But the age of computors does not alter nursing practice. All patients have vital signs taken three times in eight hours. We documented the changes produced by such simple minimal contact, a nurse taking a patient's pulse, on heart rate and the frequency of arrythmia (Lynch et al., 1974a; 1974b; 1977). Outside of such relatively aseptic situations the study of touch is difficult because touch is an intimate gesture. We are emerging from a period in which all touch was considered a kind of foreplay to sexual intercourse. I will, however, remind you that the term "heavy petting" was used a number of years ago to describe foreplay that did not terminate in intercourse. Even though we now recognize the functions touch serves in our lives apart from the prelude to genital arousal, it is still difficult to study directly. I think subjects have more problems being simply affectionate in experimental situations that they do having sexual intercourse in experimental situations.

We studied the structure of tactile communication between people and dogs by observing clients waiting with their pet in the waiting room of our veterinary clinic (Katcher, 1979). The situation is a part of the life of both dog and master. Both are habituated to waiting for their doctors, both are somewhat anxious, and the situation is defined as public.

Our first task was to observe the interactions and define a dictionary of human gestures, a qualitative description of the possible kinds of touch in this situation. This dictionary would permit us to do quantitative research at a later date. We made the decision at the start to pay attention to the person rather than attempt, at this time, to precisely describe the interaction; that is, to provide a choreographic description of the interaction. We feel, however, that this kind of interactive description will be necessary at a later date.

The understanding of the complexity of the relationship between people and their pets will not be complete without such an interactive description. That more complex task will require film rather than a human observer.

Our observer was a student trained in ethological technique. Ms. Goodman went to the waiting room with a "cover dog," Emily, a well behaved golden retriever, and a pile of school books and a note pad. At first she just recorded observations and discussed them with me to derive the basic elements of the dictionary. I should say at this time that the dictionary included positions assumed by the animal that brought him in passive contact with his owner. After we had established a comprehensive list of contacts, she made timed observations of people selected in true random fashion. These observations were used to make sure that we were not systematically overlooking certain kinds of interactions. The timed observations also provided quantitative data about the frequency of interaction between dog and person and the frequency of different kinds of gestures.

I would just like to mention three salient findings that I believe are quite important for the understanding of human behavior. The first can be simply described. *There was no difference in frequency, amount, and kind of touching between men and women.* Men touched, fondled, and caressed their animals in the same way and with the same frequency as women. This negative result has, I think, singular importance for the understanding of the meaning of dogs within our psychic economy. Men are consistently described as using touch as a medium of expression less frequently than women (Goffman, 1976). Men are believed to be less willing to demonstrate physical affection in public, to have less "need" of such affection in nonsexual contexts. In the United States men do not kiss on greeting, they do not walk with arms around each other, they are less likely to stroke or fondle children in public than women, and they are said to require less affection in intimate relationships. Women are much more likely to complain at the lack of affectionate touch in intimate relationships.

TABLE 3-VI

COMPARISON OF MEN AND WOMEN INTERACTING WITH THEIR PETS

Behavior	Women	Men
	N = 74	N = 67
Physical contact	66 (89%)	59 (88%)
Idle fingering	22 (30%)	26 (39%)
Scratching	23 (31%)	27 (40%)
Massaging	13 (18%)	14 (21%)
Stroking	26 (35%)	21 (31%)
Position contact	50 (68%)	46 (69%)
Patting	17 (23%)	16 (24%)
% Observed time spent in any form of contact	N = 37 19%	N = 23 35%

Apparently dogs are a means through which males can both express and receive affection in public situations. It may well be that for some men the only outlet for intimacy expressed through touch may be an animal. Certainly during adolescence when boys have not yet assumed a sexual role and they are rigorous in their rejection of touching affection from their parents, the presence of a dog may be their only outlet for such affection. I should add that we did not have enough data on cats to make a comparison. There is a cultural bias against male ownership of cats. In the Toronto survey (Christensen, 1979), over 90 percent of the cats in apartments were owned by women. I do not know why it is permissible for men to own and caress a dog but inadmissible for them to own and caress a cat.

Size of Dog

The second quantitative finding of importance was the absence of a difference in frequency and kind of contact when owners with different sizes of dogs were compared. For this analysis we excluded from the qualitative analysis postures that were only possible with a small dog.

TABLE 3-VII

COMPARISONS OF INTERACTIONS WITH DIFFERENT SIZE DOGS

Behavior	Small	Medium	Large
	N = 42	N = 42	N = 57
Physical contact	39 (93%)	38 (91%)	48 (81%)
Idle fingering	16 (38%)	18 (43%)	14 (25%)
Scratching	18 (43%)	15 (36%)	17 (30%)
Massaging	4 (10%)	13 (31%)	10 (18%)
Stroking	13 (31%)	21 (50%)	13 (23%)
Position contact	33 (79%)	28 (67%)	35 (61%)
Patting	7 (17%)	15 (36%)	11 (19%)

Perhaps this latter finding is an artifact of studying a situation in which people and animals are in need of frequent reassurance and the animal is in need of restraint. However, it does suggest that all breeds of dogs provide an outlet for touch and that the trend toward larger dogs is not an indication that there is less of a need to handle or fondle dogs as opposed to play with them. It also indicates that the stereotype of a women using a small lap dog for an outlet for affection or perhaps a replacement for a young child is only an invidious sexist stereotype. The pipe smoking male with his retriever is just as likely to touch, fondle, kiss, or hug his animal.

Idle Play

The third finding that I would like to discuss is perhaps more important for the understanding of the human need for touch than the two quantitative results just described. We have identified a type of touch with animals that Ms. Goodman called "idle play" and that we have also called distracted touching, petting out of focal awareness, or caressing in revery. In this style of caress, which had a relatively high frequency even in this unrelaxed and arousing situation, the person and the animal do not have each other's mutual attention. There is usually no eye contact, the person is not looking at what he is doing, and the dog is usually not soliciting the interaction. Frequently the person is obviously doing something else—talking to another person, reading, or looking about the room. Sometimes the observer noted that the subject's attention or gaze seemed to be nowhere as if he or she were inattentive to the environment or in a daze or revery.

The idle play or distracted petting could take the form of stroking, patting, scratching, ruffling the fur, smoothing the fur, rolling the fur between thumb or forefinger, tracing a pattern on the fur with the index finger, or even a plucking or grooming movement. This is the kind of fondling we think of when we picture someone in front of a fire with a cat in his or her lap, or a dog at the side of the chair. The same kind of comforting touch that puts the cat or dog into a state of inattentive comfort puts the person into a similar relaxed, unaroused, slightly inattentive state. It is very much like the relaxation we experience when looking into a fire. And how frequently in our minds are cats and dogs and fireplaces linked? Perhaps because both permit the same kind of public revery, and both are external stimuli that help us lose or lower our state of arousal. The idle stroking and fondling of worry beads by men in the Near East, where dogs (and Americans) are despised animals, is another example of the same kind of inattentive comforting by touch.

A Touch of Security

I want to leave the data on observations of people in a veterinary waiting room and examine that same kind of idle play or inattentive touch to say something more about the bond between people and their companion animals. The time at which we see this kind of idle play in its most florid state is in the lives of children under two years of age. We can watch infants use their thumb, their hair, a teddy bear, their mother, their mother's hair, their father's hair (or beard), a blanket, or any other kind of soft object to stroke, fondle, and pet when they are frightened or lonely and are retreating from the world a bit, when they are going to sleep, when they are emerging from sleep and do not wish to plunge into wakefulness, or when for a thousand and one reasons they want comfort. Looking at infants taking this kind of comfort, one is frequently struck by their state of revery—even when they are not falling asleep. They do not seem to be focusing their attention anywhere—they are lost for a moment in the sensation of touch.

The person who is holding the infant obtains some of that kind of comfort as well. Animals may be good to think but babies are good to hold. Some infants like touch more than others, and when one holds a child who is a bit low on the activity scale, one can experience the same kind of revery while idly stroking the infant, ruffling its hair, or patting it gently on the back. Sometimes one can see a parent and child, both stroking or fondling each other, both lost in their own private revery, and both in some way unaware of what the other is doing.

When the infant is put to bed in most middle-class homes, he has a teddy bear or a favorite blanket, but not another person. Not a brother, not a sister, not a father or a mother. He is taught to go to sleep by taking comfort from objects, not the warmth, touch, or regular breathing of another human being.

Now what happens when that infant starts to grow up and attains the mature age of five, six, seven, or eight? One of the first signs of growing up, in the United States at least, is an active refusal to be restrained, groomed, and touched. Children no longer wish, some of them, to be fussed over, kissed, petted, or stroked. They no longer wish to rest quietly on a parent's knee—*when the parent wants to have them on his or her knee.* At the same time they stop using their parents as objects to stroke and touch and are actively discouraged from using their thumb, fingers, hair, or any object like a teddy bear for comfort and touch.

Now I would like to suggest to you, as a testable hypothesis, that it is the age at which children begin to refuse affection with restraint, refuse to be used for idle touch, and are discouraged from using others and other objects for such comfort, that parents think of obtaining a pet. For the parents it provides a being defined as a child. Remember most people we asked think that pets are members of the family, and the family role named most frequently is that of a child under ten years of age. But it is a child that one can restrain for petting at will, train to like such restraint, and train to actively solicit affection at a high frequency and almost never refuse such affection. For the child, the animal renews the joys of touch without submission to the will of parents. With the pet, the child can set the rules, or most of the rules, for such affectionate displays.

The way in which children use pets for idle play was forcefully brought to my attention by the superb film of Dr. Ange Condoret which records the events that follow the introduction of an animal into a nursery school. In that film there is a marvelous sequence in which a child engaged in rough and tumble play with a dog gets astride of the animal, bends over, places his cheek against the dog's fur, puts his fingers in his mouth, and has a few seconds of bliss before resuming his play. One of the tasks we have in front of us is to study the connection between play and comforting touch. In play with an animal, one can alternate rough and tumble play with stroking, petting, and fondling. There are fewer opportunities for such alternation in the usual gamut of sports we teach our children.

CONCLUSION

I would like to return to the problem of research into the nature of the human companion animal bond. Relationships with animals must be studied in their own right, not as a substitute or an analogue with relationships of other human beings. The child begins a life with objects from the first moment he comes in contact with bedding rather than flesh, from the first moment he is given glucose in the nursery. He continues a life with objects before he can recognize the distinction between human beings and other animals or even between a stuffed toy or a living person. We are indebted to Searles (1960) for insisting on the importance of the natural or the nonhuman environment. The literature about pets has been encumbered by generalizations without data base, which define contact with animals as a kind of inferior substitute for contact with human beings. I am thinking in particular of the chapter on pets in Desmond Morris' book *Intimate Behavior* (Morris, 1973) and Szasz's book *Petishism* (1971). In both works the relationship with pets is considered a perversion of human relationships or a substitute for human relationships used by people who can not obtain the real thing. I think that neither of these generalizations are true. Quite to the contrary, I think we have presented evidence suggesting that the companion animal bond must be looked upon as a kind of relationship that supplements and augments human relationships—the bond distinctively different from human relationships. And those distinctions are essential to the nature of our bond to companion animals.

REFERENCES

Adel-Bath, M., Krook, A. C., Sandquist, G., and Skantze, K.: *Do We Need Dogs? A Study of Dog's Social Significance to Man.* Publication of the University of Gothenburg School of Social Work and Public Administration, 1978.

Anderson, R. S. (Ed.): *Pet Animals and Society.* London, Bailliere Tindall, 1975.

Bowlby, J.: *Attachment and Loss.* Honolulu, The Hogarth Press, 1973.

Cannon, W. B.: *Bodily Changes in Pain, Hunger, Fear, and Rage.* New York, D. Appleton and Company, 1915.

Christensen, A.: *A City of Toronto Pet Ownership Survey.* Publication of the Toronto Humane Society, 1979.

Condoret, A.: *L'Animal Compagnon De L'Enfant.* Paris, Fleurus, 1973.

_____ : Une nouvelle methode relationnelle au service de l'enfant: L'intervention animale modulée precoce. *Bull Acad Vet de France, 51:*471–74, 1978.

Corson, S. A., O'Leary Corson, E., and Gwynne, P.: Pet facilitated psychotherapy. Anderson, R. S. (Ed.): *Pet Animals and Society.* London, Bailliere Tindall, 1975.

Curtis, P.: Animals that care for people. *New York Times Magazine,* May 20, 1979.

Darwin, C. R.: *The Expression of the Emotions in Man and Animals.* New York, D. Appleton and Company, 1913.

Davis, S. J. M. and Valla, F. R.: Evidence for the domestication of the dog, 12,000 years ago in the Natufian of Israel. *Nature, 276*:608–10, 1978.

Descartes, R.: *Discourse De La Methode.* In Liard, Louis (Ed.): Garnier Freres, 1960.

Foucault, M.: *The Birth of the Clinic.* New York, *Random House,* 1973.

Fox, M.: Pet-owner relations. In Anderson, R. S. (Ed.): *Pet Animals and Society.* London, Bailliere Tindall, 1975.

Freud, S.: Letter to M. Boneparte. In Simitis-Grubrich, I. (Ed.): *Sigmund Freud.* New York, Harcourt Brace Jovanovich, 1976.

Friedmann, E., Katcher, A. H., Lynch, J. J., and Thomas, S. A.: *Pet Ownership and Survival After Coronary Heart Disease.* Presented at the 2nd Canadian Symposium on Pets and Society, May 30–June 1, 1979a.

Friedmann E., Katcher A. H., Meislich D., and Goodman M.: Physiological response of people to petting their pets. *American Zoologist, 19,* (Abstr), in press, 1979b.

Friedmann, E., Thomas, S. A., Katcher, A. H., and Noctor, M.: Pet ownership and coronary heart disease patient survival. *Circulation, 58*:II–168 (Abstr), 1978.

Goffman, E.: Gender advertisements. *Studies in the Anthropology of Visual Communication, 3*:65–154, 1976.

Group for the Study of the Human—Companion Animal Bond Newsletter, *1:* 1, 1979.

Harlow, H. F.: Love in infant monkeys. *Scientific American, 200*:40, 68–74, June 1959.

Hirsh-Pasek, K. A., and Treiman, G.: *To the Dogs: is Baby Talk Just for Babies?* Unpublished Manuscript.

Katcher, A. H., Brightman, V. J., Luborsky, L., and Ship, I. I.: Prediction of the incidence of recurrent herpes labialis and systemic illness from psychological measurements. *Journal of Dental Research, 51*:49–58, 1973.

Katcher, A. H., and Friedmann, E.: Potential health value of pet ownership. In Johnston, D. (Ed.): *The Compendium on Continuing Education for the Small Animal Practitioner.* 1980.

Katcher, A. H., Goodman, L., and Friedmann, E.: Human—pet interaction. *American Zoologist 19,* (Abstr), in press, 1979.

Katcher, A. H., and Goodman, M.: *Dimensions of Owner's Attachment to Their Pets.* Unpublished manuscript.

Levinson, B.: *Pet-Oriented Child Psychotherapy.* Springfield, Thomas, 1969.

————: *Pets and Human Development.* Springfield, Thomas, 1972.

Luborsky, L., Todd, T. C., and Katcher, A. H.: A self administered social assets scale for predicting physical and psychological illness and health. *Journal of Psychosomatic Research, 17*:109–20, 1973.

Lynch, J. J.: *The Broken Heart: The Medical Consequences of Loneliness.* New York, Basic Books, 1977.

Lynch, J. J., Flaherty, L., Emrich, C., Mills, M. E., and Katcher, A. H.: Effects of human contact on the heart activity of curarized patients in shock trauma unit. *American Heart Journal, 88*: 160–69, 1974a.

Lynch, J., Katcher, A. H., Thomas, S. A., London, J., and Chickadone, G.: Blood pressure, talking and automatic blood pressure monitoring. Submitted to New England Journal of Medicine.

Lynch, J. and McCarthy, J. F.: Social responding in dogs: effect of person. *Conditioned Reflex, 1*:81, 1966.

Lynch, J. J., Thomas, S. A., Mills, M. E., Malinow, L., and Katcher, A. H.: Human contact and heart arrhythmia: the effects of human contact on cardiac arrhythmia in coronary care patients. *Journal of Nervous and Mental Disease, 158*: 88–99, 1974b.

Lynch, J. J., Thomas, S., Paskowitz, D., Weir, L., and Katcher, A. H.: Human contact and cardiac arrythmia in the cardiac care unit. *Psychosomatic Medicine, 39*:3, 188, 1977.

Lynch, V., Thomas, S. A., Long. J., Malinow, K., Chickadone, G., and Katcher, A. H., Human Speech and Blood Pressure, *Journal of Nervous and Mental Disease. Vol. 168,* 526–534, 1980.

Morris, D.: *Intimate Behavior.* New York, Bantam Books, p. 188, 1973.

Mugford, R. A.: The social significance of pet ownership. In Corson, S. A.: Ethology and Non-Verbal Communication in Mental Health. Pergamon Press, Elmsford, New York, 1979.

Okoniewski, L. A.: *Psychological Aspects of Man – Animal Relationships.* Unpublished Masters thesis, Hahnemann Medical College, 1978.

Rosenberg, M.: When disonance fails; on eliminating evaluation apprehension from attitude measurement. *Journal of Personality and Social Psychology, 1*:28–42, 1965.

Schneider, R.: Observations on overpopulation of dogs and cats. *Journal of the American Veterinary Medical Association, 167*:281–84, 1975.

Searles, H. F.: *Nonhuman Environment, in Normal Development and in Schizophrenia.* New York, International Universities Press, 1960.

Spitz, R.: Hospitalism. *The Psychoanalytic Study of the Child, II*:113–17, 1946.

Szasz, R.: *Petishism: Pet Cults of the Western World.* London, Hutchinson, 1971.

CHAPTER 4

DOGS AS SYMBOLS IN HUMAN DEVELOPMENT

CONSTANCE PERIN

NEIGHBORS, NEIGHBORING, AND DOGS

For the past two years or so, working from the perspectives of social and cultural anthropology, I have been examining the question of how Americans live as neighbors—middle-class Americans, living in suburbs in the several regions of a very large country. My fieldwork kept on turning up the plain fact that dogs are as important as children to the relationships between households. One person put it, "Dogs, like children, can be the glue or the solvent of the neighborhood."

Indeed, the American experience of neighboring is sharply etched by difficulties over dogs, and any discussion of neighboring ideas and practices has to take them, the dogs and the difficulties, into account. Studies of kinship and family do not, however, acknowledge these nonhumans as so significant for family structure (another child? sibling? in-law?); neither do community sociologists track their importance for relationships between households, for the numbers of dog-owning households are large. Just about half of 80 million American families have at least one dog—41 million families compared to 23 million with cats. In 1978 the total dog population was estimated at about 49 million (Bush, 1978). About 68 percent of the families having a dog keep one; 21 percent own two; and 11 percent keep three or more.

These "members of the family" are of course another species entirely, and they preoccupy many disciplines quite distant from cultural anthropology. What first took my attention as an anthropologist was less the notion of the human family augmented by animals but much more the new creature, "the irresponsible dog owner." Every public issue that dogs prompt—the neighborhood frictions, traffic accidents, environmental pollution, dog bites, canine overpopulation, and abandonment—is laid at the feet of those antisocial citizens. And all agree that they are drawn from groups of all kinds—rich, poor, urban, suburban, young, old. They share one essential characteristic: they do not keep their dog on a lead or confined to their own property off the lead. They are also the people who, time after time, defeat the enactment of local "leash laws" that are often initiated by "responsible owners."

The most puzzling aspect of the behavior of these dog lovers is that they provoke harm to their dog and expose family members to the high probability that they will grieve over an accidental death or loss. The physical relationship between owner and dog, each at one end of a lead, is the most effective way to reduce any and all socially unwelcome behaviors and to prolong the dog's health and life. Yet I estimate that about half of all American dog owners eschew the lead, falling into the "irresponsible" category. If there were many fewer dog lovers of this variety we might appropriately label them "deviants" or occasional scofflaws, but the very magnitude of the public problems and public costs attributable to them puts them into the category of a dominant social force needing to be explained rather than simply and ineffectually condemned. Why and how this variety of dog-loving American has come to exist in such large numbers is the main preoccupation of this chapter.

Even so, as irresponsible as they are, these dog owners and presumably dog lovers, are no different in other respects from their more responsible brethren. So the very ideas with which our culture brings dogs into human society at all may provide clues to their behavior as well. But these will not be historical ideas because it is possible only to speculate about the origins of the domestication of the dog, a companion of humans for some 14,000 years, as best we know. I want to address instead the meanings dogs have in human life today. As a cultural relativist I put "meanings" deliberately in the plural. I suggest, for reasons that will become apparent, that American and British meanings are similar, but, not having looked into them, I have no evidence with which to make similar claims for any other western society.

Certainly I could not make such claims for any Asian society, for the fact is that what dogs mean in the West is entirely different from what they mean elsewhere. Dogs in Asia have neither the symbolic nor the real roles Westerners assign them. There, dogs are food. They are pariahs loosely attached to the edges of society. They do not work alongside farmers, herders, or hunters, and, except in Japan, they have no place at the family hearth. Yet even Japan is said to have the largest population of stray dogs of any country, a fact that might have something to do with the dog as a symbol for evil spirits in traditional Shintoism and as a vehicle for sorcery in folk beliefs. Or it might be related to peculiarly Japanese ideas about "members of the family." A contemporary Japanese anthropologist comments on current practices:

> There is a tradition of eating dog, not only in the Pacific Islands but in most of the countries of east Asia, including China, Japan and Korea. In areas where pastoralism or hunting has not been developed, the idea of a dog having a particularly close relationship to man ("man's best friend") like the Western European sheep-dog or hunting-dog, simply doesn't exist. And looked at through alien eyes, our attitudes can only seem to be a kind of fetishism. . . . For people who put more trust in dogs than in people and who have become accustomed to treating them as members of the family, the idea of eating dog is the same as eating human flesh (cannibalism). I can also recognise that it is something that can be considered disgusting. (Ishige, 1979)

I will be trying to show how Americans' dogs are creatures of culture just as much here as they are elsewhere, on the premise that only a specifically cultural study can explain the form the dog–human relationship takes in any society.

By and large, the public policy recommendations of knowledgeable researchers and humane society officials depend simply on the irresponsible owner becoming responsible. They keep repeating the very same nostrums now so unsatisfactory in ameliorating these widespread problems. Some recommendations are borrowed from other cultures, patently inapplicable to the American case. It seems that Iceland, finding its dog population disease-ridden, imposed taxes so punitive as to discourage dog ownership altogether. Dogs are barred from Chinese cities, hardly a feat in a hungry society that includes them in its food system. In Prague the dog license is free when a certificate from an obedience school is presented; otherwise, it costs eighty-five dollars.

As a result of his research in Baltimore, New York, and St. Louis, Alan Beck put the problem squarely, as much the case today as in 1974:

> The major dog problem now facing U.S. cities is not an animal surplus, but a surplus of poorly supervised pets. This is rapidly generating a situation where urban people are dividing into two warring camps, those who love dogs and those who hate those who love dogs. Public funds are being diverted from other social programs to monitor dog bites, maintain massive rabies surveillance programs, capture and destroy unwanted straying pets, replant trees, collect dead animals from the streets and respond to complaints against irresponsible animal owners.
>
> Stray or feral dogs are not as important a problem in the U.S. as straying pets. Free-ranging pets are responsible for the vast majority of dog bites, including killings of people, and dog attacks on wildlife and livestock. The myth of the "stray" dog is important in the U.S. for it displaces responsibility on wild instinct rather than faulty breeding or irresponsible owners. Even when dog rabies was a major problem in the U.S. up to the early 1950s, 75% of all exposures were from pets, not strays. (Beck, 1974)

TWO BREEDS OF DOG OWNER: RESPONSIBLE AND NEGLIGENT

Everything the negligent dog owner doesn't do, responsible owners have found to be entirely possible. Responsible owners train them—all breeds of dogs—not to bark when left alone, indoors and out. They train them to come when called, to sit, to heel, to stay. They teach them to use newspapers for defecation and urination when the weather conspires against the daily outing. On a walk, dog and owner each at an end of a lead, they position their dogs to deposit their waste where it will least harm other people's lawns and trees, and they keep public paths and parks free of feces by picking them up and putting them into the trash. If their dog is not on a lead, it is within calling distance, reliably obedient. They protect the dogs of their neighbors from the surprise attack any dog is capable of mounting. They protect their own dogs from their dashing and straying impulses—impulses that combined with their poor eyesight can get them maimed, killed, or lost. They protect neighborhood children from being traumatized by a bite and older people from being jumped upon.

Responsible owners keep track of garbage collection days and put themselves on special alert to their dog's temptations. They tenderly surrender their dogs to a humane death when that time comes, whether because their pet is sick or can no longer be kept. They do not drop them off on a country road or in a different neighborhood, abandoning them to their own limited wits. Their dogs mate at human discretion and not their species', which is, as the statistics on unadopted puppies reminds us, nonexistent. In all, there is nothing inherent in dogs or in people that makes their mutually antisocial behavior inevitable.

Despite widespread bewilderment over and disapproval of the negligent owner, one paramount question remains to be answered, Why are there negligent owners in such large numbers? Of all dog owners in a Pet Food Institute survey, 55 percent did not see any problem with loose dogs (Wilbur, 1976). I take that to provide a rough estimate of so-called irresponsible dog owners—and that amounts to about 22 million American families. The plain fact is that only if local leash laws are enforced to their limit can this large group be deterred, and no city or town can afford financially the additional policing an all-out effort would require. Nor emotionally: meter maids in New York City have refused to be deputized to enforce that city's canine waste law. "We have enough trouble dealing politely with irate overtime parkers. Please don't add the high-strung dog owner to our daily round."

These surface tensions cover deeper-lying issues. Whatever we know about those dynamics comes either from canine genetics and social behavior on the canine side or some version of an illness-cure model on the human side. That is, whatever it is that dogs mean to people, and so far the specifics of that meaning elude scientists, that undefinable something appears to be useful in treating children with psychological problems, in extending the lives of people who have had heart attacks, in providing elderly and isolated people with purpose, and in giving convalescents and the bereaved a new lease on life. Some of these claims are better demonstrated than others, but all lack a clear explanation of why any of these results should come about.

To get at these deeper-lying issues it is necessary first to explain the fact that the dog-human bond exists at all in this culture. If we were to generalize the relationship as being that of a pet-human bond, the first fact to reckon with is that dogs outnumber cats as pets two to one. This statement is complicated by the generally acknowledged fact that cats provide a far cheaper and easier caretaking pet relationship.

The route of my explanation of the dog–human bond took its direction only gradually. To clear the ground, I decided at the outset to leave unquestioned the many explanations based on dogs as a status symbol of one kind or another. The dog as property, even fashionable property, seemed to me quite unexceptional. Anything people own can be said to provide an index to their actual or hoped-for socioeconomic position. I decided to concentrate on the dog not as an item of property but as a partner in a relationship, as "a member of the family."

Nor did I think it exceptional that dogs might be said to represent "man's link to nature," as it is often put, as though we ourselves are not already there.

A next step was acknowledging that current explanations of the negligent owner's behaviors really stop at description. Negligent owners, as the humane societies, animal control specialists, and veterinarians keep on saying, are lazy, careless, and inconsiderate of others, as well as unkind to their pet, exposing it to risk of life and limb. Yet to pinpoint any specific differences from the characteristics of responsible owners is difficult. Indeed, I think it would be generally agreed upon by close observers that both come in all guises. They can be found in poor and posh neighborhoods. They live on farms, on large suburban lots, and in city apartments. They belong to the same social clubs, churches, and political parties. They keep breeds of dogs inappropriate to their particular environment—rural dogs that worry and kill livestock, suburban dogs unable to learn not to use the neighborhood's lawns and gardens, and city dogs that bark nervously day and night.

DOGS AND OUR AMBIVALENCE TOWARD THEM

As reprehensible and irresponsible as negligent owners are, they are nevertheless second to none in adoring their dogs. Yet, they put their dogs at great risk. Dogs worrying cattle can be killed on sight by farmers, for example, and dogs die in traffic no less in suburbs than in the city. I began to see the remarkable ambivalence that characterizes this entire domain; how we speak about dogs and treat them, even as we treasure them.

Coupled with the ambivalence is the fact of their very large numbers. Together these brought me to my next decision: to regard their "irresponsible" behaviors not as aberrations but as somehow built into the dog–human bond. I will be returning to that premise.

To begin with the negative side of the ambivalence, linguistically we would have to find many another idiom to say all the things we sum up as "dogs." A wide array of deprecatory terms is generally in common use, not only by those unambivalently antipathetic to dogs. A gimmick used each week on public television expresses the ambivalence perfectly. A truly cute dog named Spot appears and barks briefly just at the moment when two film critics, assessing the new movies, announce their choices for "Dogs of the Week."

A *dog* is, according to my *American Heritage Dictionary of the English Language*, "an uninteresting, unattractive, or unresponsive person," and in the way those critics employ it, "a hopelessly inferior product or creation." Poetically, doggerel does appear to be related to the Middle English *dogge*. Dogma has quite other origins. A *dog* is also a "contemptible, wretched fellow." Plural, it means the feet. *To go to the dogs* means to go to ruin. *To put on the dog* means to make an ostentatious display of elegance, wealth, or culture; to feign refinement; to be a phony. As an adjective it means inferior, undesirable, not genuine, as in *dog Latin*. A *dog's life* is an unhappy slavish existence; and a *dog's death* is a miserable, shameful end. A dog's place is at the social bottom.

Yet among all the writings there are about people and dogs, few puzzle about this "peculiar human ambivalence" that seems to be part of the bond itself. James Jordan, an anthropologist studying the rural South, finds that dogs are today, as they have been traditionally, significant partners in hunting, tracking, and guarding; and as companions (Jordan, 1975). An "unconventional" man is one who owns no dog. Yet Jordan observes that these people handle their dogs—just as often the subject of boasts about their unalloyed fidelity and their rare abilities in sniffing out quail, coons, and rabbits—callously, even cruelly. He corroborates what James Agee and Walker Evans saw forty years ago, reported in their classic *Let Us Now Praise Famous Men:*

> The dogs are all mongrel ruralhounds . . . they are almost alarmingly rickety. . . . dogs are never kindly touched by adults, unless they are puppies; the children play with them in the usual mixed affection and torture. The Gudgers feed Rowdy rather irregularly from their plates, seldom with a floor plate of his own. . . . Dogs, if they blunder into the way or are slow in obeying an order, are kicked hard enough to crack their ribs, and, in that manner which has inspired man to call them, in competition only with his mother, his best friend, offer their immediate apologies; the sickness or suffering in sickness or death of any animal which has no function as food or power goes almost unnoticed, though not at all unkindly so. . . . (Agee, 1966)

Jordan looks to their social setting to explain these "contradictory valuations" by southern white men. "The dog is treated as though it is useful *and* useless; the dog is referred to as symbolically valuable *and* worthless; the dog is employed as a standard of excellence *and* of baseness" (Jordan, 1975). No one in the impoverished rural South, suggests Jordan, has much in the way of food, medicine, or creature comforts, so that dogs are simply another living being embraced in the general deprivation. Nor is this harsh environment a spur in the development of men's softer side. People no less than dogs are expected to toughen up their feelings to endure. Their few possessions may be lost to fierce storms or an impatient landlord, so that being indifferent or loosely attached to abode, furnishings, clothes, and their dogs is a useful defense in so generally unpredictable a situation. Dogs as objects are undervalued so that their loss won't hurt. And finally asserting control over dogs (as over blacks, women, and children) may be one of the few powers a poor, white southern man can draw on, as Jordan sees it.

On the negative side of this ledger of love, we would certainly have to include the high rates of owners' failure to license their dogs and the high rates of abandonment of once owned dogs. Not even half of all owned dogs are licensed, so that once lost, those animals are likely never to return home. And dogs will stray—all dogs, given a rabbit to chase, a bitch in heat, or a travelling companion. Not being able to trace ownership, most pounds will offer up the dog for adoption after forty-eight hours or, at most, after one week. Then they will kill them. Because only about one-third of the strays are ever adopted from pounds, the chances are that an untraceable dog will die. When their dog's home address is not on the public record or attached to the dog's collar ot tattooed on a hindquarter, owners deny their dog's social connections and leave him to the mercies of strangers.

No American newspaper misses so much as a day of advertisements run under the "Lost Pets" columns. Notices for beloved but straying dogs on neighborhood trees and telephone poles are as much a feature of our social landscape as mailboxes. About 1,000 radio stations participate in a "Pet Patrol Radio Directory" set up by the American Humane Association to broadcast missing-pet notices — about 35,000 over the past several years. Of all former dog owners, 24 percent no longer have their pet because it was killed in an accident (18%) or ran away (6%), according to the Pet Food Institute. One of the reasons people give for not getting another dog is that they could not bear the grief its death would bring. Of the former owners, 15 percent felt its loss too keenly. Yet, aching as the loss is, millions of owners are nevertheless doing too little, somehow, to be sure of avoiding the experience.

The positive side of the ambivalence consists only of superlatives. There are no gradations to this side of the valence: people *love* dogs. They're not just fond of them. They don't like them just a bit. People *love* dogs. *I* love dogs. The superlatives abound, and topping the list is *man's best friend*. Everything about the dog's relationship to people is lavishly phrased. The defining canine trait is said to be fidelity — utter devotion, everlasting. Scientists, not mere sentimentalists, describe the dog's capacity for unconditional, indeed "immeasurable," adoration.

> The fidelity of a dog is a precious gift demanding no less binding moral responsibilities than the friendship of a human being. The bond with a true dog is as lasting as the ties of this earth can ever be. ... The plain fact that my dog loves me more than I love him is undeniable and always fills me with a certain feeling of shame. The dog is ever ready to lay down his life for me. If a lion or a tiger threatened me, Ali, Bully, Tito, Stasi, and all the others would, without a moment's hesitation, have plunged into the hopeless fight to protect my life if only for a few seconds. And I? ... Every dog that ever

followed its master. . . . : an immeasurable sum of love and fidelity. (Lorenz 1954)

"Marvellous as may be the power of my dog to understand my moods, deathless as is his affection and fidelity, his mental state is as unsolved a mystery to me as it was to my remotest ancestor" (William James quoted in Papashvily, 1954).

SUPERABUNDANT LOVE

The quality that best characterizes the bond of feeling between people and dogs is abundance; in fact, superabundance. For the bond is often seen to represent an *excess of love* having no rightful place in human relationships, supersaturated feelings people are not able to or not allowed to bestow on other people. In describing their opponents' arguments, Town of Wellesley people favoring a leash law said they "rhapsodize" about their dogs. People were found to own pets for the "satisfaction of giving and receiving complete and total love and devotion." George Bernard Shaw, ever doubtful, put it that animals in general "bear more than their natural burden of human love." Ambrose Bierce's definition from his *Enlarged Devil's Dictionary*: "Dog, n. A kind of additional or subsidiary Deity designed to catch the overflow and surplus of the world's worship. . . . "

Not only abundant, but the more powerful for being visceral: dogs enter into our deepest feelings without our having to use speech. Lorenz had one dog that "reacted to any symptoms of illness, and expressed her anxiety not only when I had a headache or a chill but also when I was feeling downhearted. She would demonstrate her sympathy by a less cheerful gait than usual, and with subdued demeanour would keep strictly to heel, gazing up at me continually" (Lorenz, 1954).

The superlatives translate into idealization. In interviews with municipal animal officers and with suburbanites, over and over I was told that people with bothersome dogs will deny their neighbor's complaints with far greater vigor than they would defend their children's misbehaviors. "My dog would never. . . . " bark, dig, or defecate on that lawn, or tip over the garbage can and nose through it for a snack. These denials are so blanket and so expected that animal control officers have a fixed routine for investigating neighbors' complaints first hand and over a period of time to put themselves in the best position for countering such denials.

I've observed many a defensive posture in dog owners. The words say that their dog is perfect, but their body language, as they shift about, clear their throats, avoid my eyes, and thrust forward their chins, belies their assertions. Dog owners and nonowners differ considerably in their perceptions of the problems that dogs cause. While 38 percent of former owners and 46 percent of people who have never owned a dog see the problem of animal waste deposits, only 22 percent of present owners do. Although 18 percent of nonowners acknowledge the problem of dogs' attacking or frightening children, only 9 percent of the owners recognize it. Just 45 percent of the owners see a problem with animals running loose, compared to 36 percent of former owners and 28 percent of those who have never owned a dog (Wilbur, 1975).

The idealization of the dog is, then, the first characteristic of the relationship. The second is a belief that, as Desmond Morris puts it in no uncertain terms, dogs are "like children" to their owners. "Virtually all pet-keeping can be seen. . . . as a form of pseudo-parentalism, with the animals standing in for missing infants, either because the 'parents' are too young to have real children, or because for some reason they have had children but are now without them. These are the conditions in which pet-keeping is at its most intense" (Morris 1977).

Indeed, I had thought that dog owners were defensive because they took criticisms of their dogs as criticisms of themselves. But shouldn't that sort of defense as a general rule extend first to their own flesh and blood, their actual children? Would they expose their children to life-threatening risks? Would they allow them to act antisocially without calling them on it? To be sure, to all appearances dogs seem to be treated as we treat children (and all those less able to fend for themselves in our terms). The analogy to children seems apt.

Still, many dog devotees do not infantilize their pets or the species. Thomas Mann, James Thurber, and Helen and George Papashvily, write with affection and with great respect for their dogs' admirable traits, never having recourse to an idiom of infancy or dependency. Konrad Lorenz's popular book, *Man Meets Dog*, gave that perception perhaps its widest currency. He asserts authoritatively that the lifelong neonate or childlike qualities of the dog account for its human appeal; misreading, I think, the significance of submissive gestures as I will explain shortly.

The Parental Bond

I kept being struck nevertheless by the common observation that dog owners are *more* defensive about their dogs than about their children, evidence I took to suggest that dogs and children might not be conceptually equivalent categories. And so I decided that there was sufficient reason to work on the doubt that dogs are surrogates for children; that it was possible the relationship with the dog might mean something else; that aside from the caretaking analogy with children a symbolic relationship stood in the shadows of the real relationship, one consisting of all those things dogs and owners do with and for one another: take walks, listen to monologues, chase balls, guard the house, herd sheep, and just be there when we come home, stirring up, as Thoreau put it, the dead air in a room. The symbolic relationship might clarify the nature of the irresponsible owner's behavior.

My conclusion is that there is indeed a parental bond but one that I propose takes the opposite form. We are, speaking symbolically, the children of our dogs. Our species difference further signifies that ultimate yielding of our parental ties and, in growing up, our coming to terms with our separateness. The Anglo-American bond with dogs is, I will try to show, a symbol of the most fundamental properties of human existence as our culture has come to understand it.

Given the fact that the dog and not another species is the vehicle for these meanings, then how dogs see the bond must contribute to any meanings in equal measure. In 1965 the zoologists John Paul Scott and J. L. Fuller published the results of fifteen years of experiments in dog breeding and social behavior. They found breed differences in behaviors of various sorts, but, by and large, their evidence leads to the general conclusions I report here.

The "one-man dog," they found, is something of a myth. Dogs respond equally to different handlers. A test of the belief that the person feeding the dog receives the greatest amount of the animal's loyalty showed that feeding is not necessary to the development of a social tie. Even if people never feed them or act passively toward them or punish them for making friendly overtures, dogs will become closely attached (Scott et al., 1965). "A puppy does not automatically love you because you feed it," they conclude.

Canine socialization differs considerably from ours. A series of experiments revealed that when the puppy is taken early from its litter (at about three weeks) it forms paramount relationships with people, not with a single person. If taken from the litter at a later time, it forms strong relationships with both people and dogs. And if taken from its littermates still later (after about fourteen weeks), it has already formed strong relations with dogs and its ties with people will be relatively weak. By contrast, humans' strongest bond, Scott and Fuller point out, is between older and younger individuals, adults and their young. But the dog's most significant relationships are with contemporaries. The dog's relationships with people mirror their relationships to littermates who, in the wild, form a cooperative hunting pack whose work is to stalk food for all. The "submissiveness" of dogs below the "top dog" signals more their location and cooperative role in this food-getting system than it does "being dominated." Still, that aspect certainly exists, and in also having a "dominance hierarchy" the human family provides a parallel to the sort of group dogs are equipped to relate to. In the "good family dog" we recognize that biological basis for the two species coming together.

Dogs are genetically constituted then, less to ask us for mothering or parenting and more for the companionship of colleagues and the sharing of tasks. "A well-developed leader-follower relationship is not a characteristic of either dog or wolf societies" (Scott and Fuller, 1975). That is not to say that dogs may not be trained, wittingly or not, to fawn over a master or mistress or to be very dependant all their lives. But their natural capacities are otherwise. Dogs are not constituted to be "utterly devoted" to any one of us, in plain fact.

We have that interest. Dogs keep on telling us that our definition of "fidelity" is not theirs when they so readily take to a new family or wander away without so much as a tear. The "one-person dog" is our myth (an observation dog haters often make). These are *human* ideas defining the dog's relationships to us. They are cultural ideas.

HUMAN EXPERIENCE AND THE IDEALIZATION OF THE DOG

Turning to our own species, I wondered what else in human experience might our idealization of the dog resemble or recapture. When else have we ever actually received unquestioning devotion, utter adoration, a total absence of judging, unspeakably overwhelming trust, unspoken understanding, and unbounded love? How does it come to be that for our "best friend" we turn to another species? That the fulfillment of such supersaturated expectations may be anticipated only from another species brought the realization that I was in the presence of something transhuman or metaphysical—that is, a symbol, a condensation of meanings richer than real.

A line from Walker Evans and James Agee began to resonate, "that man has been inspired to call dogs in competition only with his mother, his best friend. . . . " I began to look into the nature of the mother-child bond thinking that those feelings of idealization, of ambivalence, of complete devotion might take their strength, their force, from this source.

That dogs serve as "transitional objects," in a fashion similar to teddy bears, security blankets, and any one of a number of the soft talismen that youngsters carry around to provide comfort when they have been disappointed or are feeling lonely, has been observed by some psychologists and psychiatrists. They do not reconcile that, however, with their equal certainty that dogs represent surrogate children. Although they do not say so, perhaps what they mean is that babying encompasses being babied in that people are giving what they want to get, or that in the minds of infants they and their mother are felt to be one. But there is an important difference between teddy bears and dogs. Dogs are not inanimate but living creatures with whom we have distinctive relationships as much shaped by their species characteristics and individual makeup as by ours. The teddy bear's responses are as we imagine them to be, but Rover's are really those of a dog.

Whatever the meaning of any kind of transitional object, the important thing is that the maternal closeness and the warmth it represents speak only to the positive side of that bond. The time comes for infants to be weaned. Not much later other rudiments of the infant's independent existence begin to appear—talking to make needs known rather than being anticipated, then walking. And soon infants are satisfying their own curiosities, gradually able to leave mother and father farther and farther behind. Becoming a separate person, that is the Anglo-American way. But it is also the agony and the key to the meaning dogs have for us.

The mother-child bond through our earliest years is composed of three main elements which, reaching a balance with one another, become the tripod of our psychic organization: Our visceral feelings of comfort and intimacy, our feelings of self-confidence and autonomy, and our feelings of disappointment and loss at having to give up some of the first to experience the second. We work out this balance with the help of our mothers and fathers who, in the measure that they in turn have it, can love us even while our disappointment generates anger at them. Our culture defines the attenuation of that symbiotic relationship as the beginning of "growing up," on the way to becoming emotionally mature and ready to found families of our own. The bond does not pose any problems, only the breaking of it.

Loosening it is not a once-and-for-all event in our lives. As we experience many other kinds of relationships we realize that the original counterpoint of pleasure and pain has turned into a permanent octave of affect we keep replaying in different keys. In its minor mode, psychoanalysts refer to the end of our "merger with mother" as "the depressive position," each similar experience reawakening us to our inevitable separateness from one another. But I want to stress it is not simply an aching over a loss. It is compounded further with our disappointment and anger that the break happened at all. Because infants identify with their mothers, they tend to turn the anger upon themselves: the yeast of depression throughout our lives. Lorenz recounts that his dog went into a depression upon being left behind. He does not speak of his own.

Among students of the mother-infant relationship, there appears to be an informed consensus about the character and structure of the earliest months of human life in American and British societies. An American psychoanalyst, Margaret Mahler, defines three phases of the separation-individuation process (1972). It begins only after the "peak of symbiosis" has been experienced, at about four to five months. Then in that first period of "differentiation" the child recognizes the mother as distinctly another being. That phases into a "practicing period," lasting from about seven to ten months up to fifteen or sixteen months, when the child's physical development permits it to actually move away, first by crawling and climbing, followed by the elation of walking alone, the drunken joy of the toddler. Mahler calls the third phase "rapprochement," when the child also wants to share everything with its mother. Although more mobile than ever before, the child keeps touching base, seeking "constant interaction" with the mother, and with father and other adults as well. As the toddler approaches twenty-five months, the exhilaration of the practicing period begins to wear off, and the child begins to realize that "the world is *not* his (sic) oyster; that he must cope with it more or less 'on his own,' very often as a relatively helpless, small and separate individual, unable to command relief or assistance merely by feeling the need for them, or giving voice to that need" (Mahler, 1972). The tug between remaining in close touch (constantly demanding the mother's attentions) and moving off to take in the world as a separate being is not only the toddler's crisis but is, Mahler suggests, "the mainspring of man's eternal struggle against both fusion and isolation." "One could regard the entire life cycle as constituting a more or less successful process of distancing from and introjection of the lost symbiotic mother, an eternal longing for the actual or fantasied 'ideal state of self,' with the latter standing for a symbiotic fusion with the 'all good' symbiotic mother, who was at one time part of the self in a blissful state of well-being."

D. W. Winnicott, a British psychoanalyst and author of *The Maturational Processes and the Facilitating Environment*, proposes a complementary view of the separation-individuation process. He suggests that when the peak of symbiosis has ended ("merger with mother" is his phrase), the "mother tends to change in her attitude.

It is as if she now realizes that the infant no longer expects the condition in which there is an almost magical understanding of need. The mother seems to know that the infant has the new capacity of giving a signal so that she can be guided toward meeting the infant's needs." Subsequent to this phase, "in which the infant depends on maternal care that is based on maternal empathy," contributing profoundly to the child's ego development, is the "differentiation into a separate personal self."

All told, the character and structure of the earliest human relationship reappear in the meanings dogs hold, generating the symbolic side of our real relationships with them. The symbiosis characterized by empathy and nurture flowing from mother to infant, the mother's quick responses to the child's active signalling, the child's evolution of independent mobility and a more verbal vocabulary of demands, the child's conflict between staying close and going away—not any of these particular phases but the unique tone of that earliest and singular time is recreated in dogs' symbolic relationship with us. The symbol reaches across species because we may not recreate it with any other person. For when people do try to carry much that same tone into their real relationships with other people, it becomes a kind of hum, an undercurrent that turns into static in intimate relationships, perhaps their most prevalent source of interference.

It is exactly the recognition that there is no possibility of duplicating the structure of that first relationship that fills out the meanings of the human bond with dogs. To be human carries with it the knowledge that we are separate throughout life and small in the scheme of things.

"Separation anxiety" labels those feelings experienced in infancy, early childhood, adolescence, and in those adult bereavements that accompany losses (and even growth) of many kinds. The deaths of relatives and friends, but also changes of job, of residence, and of definitions of ourselves, can cause separation anxiety. But it is not just of the fear of going it alone. Nor is the open-ended quest for some semblance of our parents' certain embrace the hardest part. "So long as a child is in the unchallenged presence of a principal attachment-figure, or within easy reach," according to the observations of John Bowlby, "he feels secure. A threat of loss creates anxiety, and actual loss sorrow; both, moreover, are likely to arouse anger" (1969).

More difficult to deal with is the anger that follows on our disappointment that the steady comforts do not go on and on. Even as those earliest pleasures are remembered, they are also suffused with memories of the rage. Although normal and expectable, the axioms of our culture enjoin us to master signs of it. Better to deny it with a calm facade; let the depression it fosters be a private sorrow.

Our parents' capacity for helping us to convert that infantile anger, raw and unprocessed, into negative feelings appropriate to the person, place, and time tests them perhaps more than any other task in our growing up. It depends on their ability to see the "terrible two's" and childhood outbursts of abuse of them, of others, as temporary, not as a reason to hold a grudge. The better they do, the better able we become to relinquish its intensity because the more that parents react to it, the more they may foster infantile fantasies of omnipotence. If confirmed in the idea that these rages matter greatly, that all out of proportion to one's size and strength one can hurt deeply, the child comes to believe that any expression of these feelings will have the effect they fantasize and, with infant logic, magnify.

The "anxiety" of separating is exaggerated in infancy, as is almost everything about the world as children know it. Parents are all-powerful and perfect. They imagine themselves to be the total center of their parents' lives. And their demands, no matter how overwhelming, are met. Separation evokes an infant's exaggerated sense of helplessness and grief. The "uncertainty about the emotional availability of the parents," according to Mahler, creates a "hostile dependency upon and ambivalence toward the parents." Infant temper tantrums and the negative spells of childhood express that fury. If identified with the parents, this disappointment and anger can turn inward, and that harm to self-esteem is the harbinger of depression. The other events going on in a child's life around that same time — mother's next pregnancy, the birth of a sibling, the beginning of the rewards and punishments of toilet training — compound these uncertainties about being wholly the center, wholly loved.

The stability of our psychic tripod—that balance among our feelings of intimacy, autonomy, and anger—is challenged in every human relationship. Love is woven from these strands. Each pulling at the other provides its strength. When our hearts are broken that tension essential to love has given way. For when we experience the real loss of a beloved—a spouse dies, a friend moves away—we sink into grief not just because the ballast provided by the real closeness is gone. The ache of loss is compounded by anger at the deprivation itself. Being hesitant to enter at all into a relationship may stem from the fear that the anger will preempt more than its fair share. It is this balancing act that is at the heart of loving. Once beyond infancy, the way we juggle among these contradictory feelings has much to do with living contentedly with ourselves and living tranquilly with relatives and friends and neighbors.

Throughout life that octave of affect resonates and sometimes rumbles. One study of hospital admissions for physical ailments alone (not for psychiatric or surgical help) found "a remarkably high incidence of major interpersonal separation and loss accompanied by feelings of helplessness and hopelessness—occurring shortly before onset of the symptoms leading to hospitalization. A variety of medical disorders were involved, not just the ones usually regarded as psychosomatic disorders" (Hamburg, 1963). Even people asked simply to discuss the possibility of losing affection and respect from people close to them (and in an experimental setting) exhibit symptoms of anxiety, anger, and depression.

Feeling better doesn't mean restoring the earliest, deeply pleasurable feelings of being at one with mother as such. Mostly we replenish that early joy by loving others and being loved back.

INTIMACY, AUTONOMY, ANGER—AND THE DOG

Separating and individuating is, then, the developmental process through which we might explain the widespread ambivalence toward pet dogs, their idealization, and the denial of their ordinariness. Above all, it is the theme or structure around which dogs are shaped as the symbolic vehicle of that excess of human love, an *idea* about love apart from any real person, for the superabundance of love after infancy has no rightful human object in our society. Yet the original symbiosis is recollected. Dogs give "complete and total love," "utter devotion," "lifelong fidelity;" the "one-person" dog. Speechless, yet communicating perfectly, the mute and ever-attentive dog is a symbol of our own memory of that magical once-in-a-lifetime bond.

But that blissful memory does not motivate the whole of this singular symbol. The fact of the species difference is fundamental to our bringing dogs into our lives. That difference symbolizes the reality of the original separation and our recognition of our ultimate apartness from mother and from one another. We become adults and we know we have awakened from that dream. Dogs, far from being an icon of infantility, symbolize the perpetual recapitulation of everyone's most fundamental, most profoundly felt struggle: to come to terms with the inevitability of the uncertainty and loss inherent in our first social relationship without losing faith in others or in oneself.

Our relationship to dogs symbolizes our own fidelity to human continuity, biological and emotional. The meanings that this symbol makes available renew people's trust in one another. They help to make society possible.

These meanings hold for all dog owners, responsible and negligent. But in behaving toward a beloved dog in ways that contradict the loving, as negligent owners do, they reveal just how much struggle maintaining a balance can be. For they are, by default, hoping that their dog will get lost or hurt or killed. Unmistakably, they put their dog at risk. Their free-roaming dog may harass neighbors, or worse, bite children on the block. Just as they are sure that their pet, perfect as it is, would never do such things, they might find it equally difficult to acknowledge that their love for their parents is a flawed compound; that it includes heartfelt anger as well. Responsible dog owners may struggle no less with this most general issue of living. They do not happen to have chosen their dog as a vehicle of this perplexity. None of us is free of it. We each work out a characteristic style for keeping our tripod in balance.

So fundamental is this ceaseless loop of closeness, separation, and disappointment, and so noticeably is it a leitmotif of our lives the whole of their length, that its appearance as a central meaning in our relationship to dogs should be no surprise, for they stand next after family and friends as companions with whom we have come, in our culture, to share ourselves.

Copyright 1981
Constance Perin

REFERENCES

Agee, James and Evans, Walker: *Let Us Now Praise Famous Men.* New York, Ballantine Books, 1966.

Beck, A. M.: The experience of U.S. cities. *Proceedings of the First Canadian Symposium on Pets and Society, An Emerging Municipal Issue.* Toronto, Ontario, Canada, June 23–25, 1974, p. 58.

Bowlby, John: *Attachment.* New York, Basic Books, 1969, p. 209.

Bush, R. R.: Pet birth control market research. In *Proceedings of the Symposium on Cheque for Canine Estrus Prevention.* Brook Lodge, Augusta, Michigan, March 13–15, 1978.

Hamburg, David: Emotions in the perspective of human evolution. In Knapp, Peter H.: (Ed.): *Expression of the Emotions in Man.* New York, International Universities Press, 1963, p. 315.

Ishige, Naomichi: Roasting dog (or a substitute) in an earth oven: an unusual method of preparation from Ponape. In Kuper, Jessica (Ed.): *The Anthropologists' Cookbook.* New York, Universe Books, 1977, pp. 204–5.

Jordan, James William: An ambivalent relationship: dog and human in the folk culture of the rural South. *Appalachian Journal, 2,* no. 3, Spring 1975.

Lorenz, Konrad: *Man Meets Dog.* London, Methuen, 1954, pp. 132, 142, 198.

Mahler, M. S.: On the first three subphases of the separation-individuation process. *International Journal of Psycho-Analysis, 53,* 333–38, 1972.

Morris, Desmond: *Manwatching: A Fieldguide to Human Behavior.* New York, Abrams, 1977, p. 265.

Papashvily, Helen and Papashvily, George: *Dogs and People.* Philadelphia, Lippincott, 1954, p. 35.

Scott, J. P. and Fuller, J. L.: *Genetics and the Social Behavior of the Dog.* Chicago, University of Chicago Press, 1965, p. 145.

Wilbur, Robert: Pets, pet ownership and animal control: Social and psychological attitudes, 1975. In *Proceedings of the National Conference on Dog and Cat Control.* Denver, Colorado, February 3–5, 1976.

Winnicott, D. W.: *The Maturational Processes and the Facilitating Environment.* New York, International Universities Press, 1965, p. 50.

CHAPTER 5

THE PSYCHOLOGICAL DETERMINANTS OF KEEPING PETS

GISELHER GUTTMANN

DIFFERENCES BETWEEN ATTITUDE AND BEHAVIOR

The assumption that human behavior is in harmony with attitude and that by changing attitude one automatically changes behavior has been seriously questioned by recent investigations (Stundner, 1977; Weingarten, 1978).

This discrepancy was illustrated initially through a study on attitudes toward safety in an Austrian Vocational School in which apprentices were prepared both theoretically and practically for various trades. A very comprehensive educational program was introduced that dealt with attitudes to safety and avoidance of accidents at work. The apprentices became familiar in the best possible way with all the relevant information, and in the course of time they learned how to recognize and avoid all risk situations. For example, always be careful to wipe your hands clean of oil remains before gas welding, otherwise when using the welding gas oxyacetylene there is danger of an explosion; or when working under a hydraulically jacked-up weight, always make use of another mechanical prop in order to avoid the weight falling if the hydraulic pipe bursts.

After completing their training, the apprentices were examined by a questionnaire on the relevant safety precautions, and it was confirmed that they had acquired the attitudes and knowledge necessary for avoiding accidents. Shortly after this, one of the previously mentioned situations was worked into the everyday routine without the apprentices being able to suspect the experimental character of the situation. The foreman gave directions that the shock absorbers of a car be changed. As soon as an apprentice had begun this job, the foreman interrupted him and asked him to weld an exhaust pipe quickly. Hardly one of the apprentices, who shortly before had proved that he possessed the relevant knowledge and had described a person as extremely careless who welded with greasy hands, cleaned his greasy hands before opening the welding bottle.

In the same way, an apprentice called over to work under an already jacked-up weight began a job that involved a high expenditure of energy (changing a part of the chassis) without taking a second mechanical precaution, etc. (We used empty welding bottles as well as a hydraulic system that was strengthened in the middle by a rigid iron bar. Otherwise, we may well have suffered losses with our own experimental models.)

This example, based on the model of safety precautions, showed us that considerable discrepancies can exist between knowledge and attitudes of mind and actual behavior, which can have marked, practical consequences.

Fishbein (1975) formulated a comprehensive mathematical model that not only enables one to grasp hidden motives but also allows one to determine to what extent the various components contribute to the creation of a willingness to behave in a certain way. This was integrated by us into a series of studies in which the previous examples of discrepancies between basic attitudes and actual way of behavior were observed.

APPLICATION TO PET OWNERSHIP

This model has now been tested by Zemanek (1979) in the field of pet keeping. Random checks have shown that the attitude toward a pet cannot determine the way of behavior. In our country at least, even nonpet owners have on the whole a remarkably positive attitude toward pets. This attitude is in many aspects so animal orientated that it is hardly distinguishable from that of the actual pet owners. The question arises as to why they do not in fact keep a pet.

By making spot checks on sixty-two pet and nonpet owners in Vienna, these facts were examined empirically. The necessary attitudes of mind that determine whether a person is actually prepared to keep a pet were registered.

A questionnaire was produced in which the experimental models had to state their opinion on all important aspects of attitude and behavior by ratings on a seven point scale. This questionnaire permitted an analysis of the effective components (evaluations), the cognitive components (beliefs), as well as the normative tendencies (normative beliefs) and a willingness to behave in a manner corresponding to that of the majority (motivation to comply).

The evaluation of the first study produced a series of definite results of which the most important will be summarized as follows:

The decision to keep a pet (or to think of keeping a pet) or not is clearly defined in all cases. On the relevant scales, pet owners and nonpet owners show clearly diverging attitudes of mind (Fig. 5-1). This discovery is not all surprising and simply shows that in our population most people are capable of living according to their motives as far as the keeping of pets is concerned and of adequately fulfilling their desires. This strong group difference only becomes surprising when we look at the direct attitude toward a pet. Do nonpet owners find animals less interesting or attractive? By no means! *Both* groups already have a remarkably positive attitude in their immediate evaluation of pets so the question arises as to why the nonpet owners, in spite of their extremely positive attitudes, are so unwilling to take on a pet? Factors that determine this behavior are not to be found in the direct attitude to animals but rather in other hidden spheres.

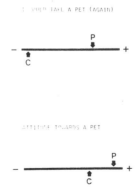

Figure 5-1. In this and the following figures, attitudes of petkeepers and controls are shown on a scale. The right end means perfect agreement, the left end disagreement. One may see that P show a high tendency to keep a pet again while C show extremely low motivation, although the source attitude does not differ so much.

PERSONALITY DIFFERENCES

Are there such things as typical pet owner and nonpet owner personalities? As further analyses have shown, there are in fact differences to be made in a whole range of attitudes of pet and nonpet owners.

The most important differences between pet and nonpet owners are shown in Figure 5-2 and can be categorized in two broad spheres of attitude. The first group of attitudes can be described as a "permanent acceptance of commitments" and toleration of possible burdens. On the whole, nonpet owners show a greater tendency to be independent and to avoid any lasting obligation. They shun the necessity of constantly having to care for something.

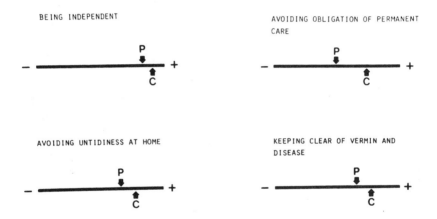

Figure 5-2. Differences in attitudes between P and C. Some attitude dimensions not related to pets are remarkably small.

One of the heaviest burdens concerns the immediate environment and the extent of its tidiness and cleanliness. Generally the nonpet owners place more weight on having a tidy and clean home. Avoidance of dirt and vermin is noticeably more important to them than to pet owners.

The second group of attitudes concerns social environment. Pet owners generally show a greater tendency to avoid loneliness or being alone. They respond to the appeal to be needed.

All the differences mentioned are clearly very small, as the sketch shows, and as can be seen from regression analysis, they are by no means in a position to account for even a small part of the discrepancy between pet and nonpet owners.

The picture changes, however, when we look at how important our experimental models rate a pet in the various named spheres (Fig. 5-3). The idea that a pet could restrict one's personal freedom does not occur to the pet owner. In contrast to the control group, the animal is far more likely to give them the most desirable feeling of being able to be on hand for someone. Through this the initially small difference in attitude becomes very large and leads to a tremendous difference in the way of behavior. The same applies to accepting the consequences that affect the immediate household environment. Pet owners are less concerned with the dangers of their home being soiled or of coming into contact with diseases than those people in the group of nonpet owners.

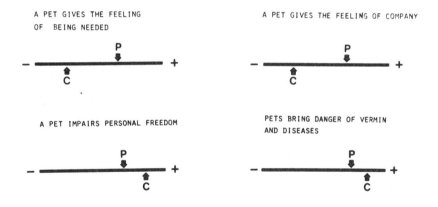

Figure 5-3.   The role of the pet in the attitude dimensions shown in Figure 5-2.

The differences are at their most evident in the sphere of the social environment (Fig. 5-4). The pet owner feels alone without a pet. An animal gives him the feeling of having a companion. The pet is really someone to whom he can talk.

In addition to this, there are a number of attitudes that stress the role of the pet as a companion of the same species. This can be

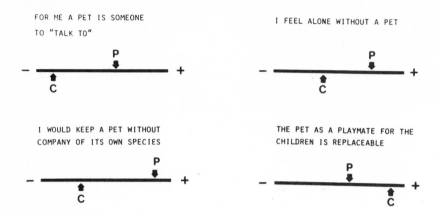

Figure 5-4.   The most significant motivation for petkeeping: the pet as partner and fellow.

seen in the fact that pet owners are more than ready to keep a pet without any companion for it of its own species. The reason for this must have deep roots that go beyond the predominant attitude. For pet owners the world of the pet is not an enclosed existence, shut off from the human sphere by a glass wall. The pet is nearly a partner of the same species, who does not need companions of its own type in order to be able to lead a happy and meaningful life.

PET OWNER ATTITUDES

In addition to this there are two attitudes from the normative sphere. All pet owners show a greater readiness to keep a pet when their circle of friends and those people whose opinion matters to them approve and agree with keeping a pet. These people would also be more willing to take in a pet that was in difficult circumstances (Fig. 5-5).

If, with the help of the mathematical model already mentioned, we try to determine which attitudes are of special importance for actually keeping a pet, then we can greatly reduce the number of relevant dimensions and deal with a very small group of attitudes that influence behavior. In doing this, it is best not to take the actual behavior as the external criterion but rather the *readiness* to behave in a certain way, which can be determined by scale tech-

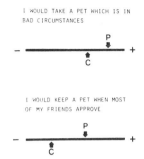

Figure 5-5.   The importance of normative beliefs.

nique. In other words, those pet owners who actually own a pet but basically show little willingness to do so should be included in the control group. The control experimental models who do not actually own a pet but who are in favor of them, show a willingness to own, but who are at present unable to own one for one reason and another, should be included in the pet owner's group.

The most important attitudes in determining willingness to behave in a certain way are (1) the feeling that the pet is something to whom one can talk and (2) an ability to regard the pet as one of the same species, which one is prepared to keep without a second pet.

The second attitude from the normative sphere is to have a positive outlook toward petkeeping itself; that is, toward all the activities and tasks connected with petkeeping. In addition, there is the attitude of the circle of friends. The opinion of the social environment is of importance for the person concerned. These individual attitudes show, by means of the actual willingness to behave in a certain manner, the remarkable multiple correlations of r to 0.81, and encompass approximately two-thirds of the real behavioral variant.

There is one interesting point to be found in the secondary results. If we compare people who have bought their pet themselves with those who received one as a present, we see remarkable differences in the structure of their attitudes. The pet buyer's attitude that a pet is a partner that gives one the feeling of being able to give something warmth and attention is to a large extent responsible for whether that person is happy to keep the pet or would be prepared to buy another one. For those people who received the pet as a present, the normative belief is the most decisive factor in determining behavior. Only when most of their friends approve of keeping pets would they also be prepared to keep a pet again in the future. The correlation between these individual attitudes is 0.82 and is, when considered alone, as important as all three part attitudes in the previously mentioned study.

FURTHER STUDIES

An attempt to understand and to look objectively at attitude components behind predominant motives is to be continued by us on a purely physiological level. It has been shown that by examining suitable characteristic values from the electroencephalogram, especially the cortical components, useful pointers can be found as to what is important for a person, to what extent they are roused to action by information, and how long this new information holds an attraction for them. These techniques will be employed in future studies.

The available data, however, may help to modify attitudes and opinions toward pet ownership. In spite of a generally positive attitude, numerous people are not prepared to keep a pet because in some particular spheres of attitude pet keeping has a specially negative-ambivalent value. If we want to make pet keeping more acceptable to them, then we must make quite clear that responsibility is not a burden. We must also bring home to them the idea that an animal is not a being from another planet but that it will be a real friend and companion.

Present development which, for the city-dweller, is creating an increasing alienation from nature and from an understanding of its laws, presents us with the task of trying to correct this attitude among nonpet owners whose low behavioral motivation lies to a certain extent in a false appraisal of secondary motives that are not related to animals. Pet keeping is more than a "hobby," it is the only effective way of maintaining an understanding of nature in an overcivilized world. How our world will be tomorrow probably depends on this attitude.

SUMMARY

Among a group of pet and nonpet owners in a large town there is a remarkably small difference between their predominant, direct attitude toward pets. Even nonpet owners are basically in favor of pet keeping. It is not necessary or thought fit to motivate them by stressing the positive aspects of a pet — their basic attitude is already positive.

On the other hand, there is a great difference in the actual behavioral tendency in which the nonpet owner differs from the pet owner. Since the basic attitude toward a pet cannot be responsible for these differences in behavior, we looked for the motives behind this difference. We observed a series of characteristic attitudes that define the groups of pet or nonpet owners. Nonpet owners show a stronger tendency to be independent and to avoid permanent ties. They place more importance on cleanliness in their immediate environment and are less disturbed by loneliness and by being alone. These differences are statistically proved but are quite minimal. Considered alone, they cannot account for the marked differences in behavior.

Behavior differences can be understood when we look at the role the animal plays for each group. For the pet owner, the pet satisfies ones need to be wanted and to be needed by someone. The dangers of a soiled home are far less important for him than for someone from the control group.

If we want to change undesirable prejudices towards pets, then it is clear that this cannot be achieved at the level of attitudes relating directly to animals, since these are noticeably positive both with pet and nonpet owners.

It would be necessary to modify the value of a pet in the group of nonpet owners into two spheres of attitude that are not directly connected with animals and relate to (1) the acceptance of responsibility and (2) social contact.

By means of suitable information it must be made clear that a duty need not necessarily be something unpleasant and that responsibilities toward a living being can become very positive.

The second sphere of attitude in which we must appeal to the nonpet owner is his silent acceptance that being indifferent to animals is inhuman, so that he sees in the animal a strange and incomprehensible world.

By means of relevant information we must make him aware that he can find a real partner in a pet, which can be an irreplaceable teacher for him in gaining an understanding of nature and her laws.

REFERENCES

Ajzen, I. and Fishbein, M.: Attitude—behavior relations: a theoretical analysis and review of empirical research. *Psychol Bull, 84*888–918, 1977.

Bergler, R. (Ed.): *Das Eindrucksdifferential.* Huber, Bern 1975.

Fishbein, M. (Ed.): *Readings in Attitude, Theory & Measurement.* Wiley and Sons, New York, 1967.

Fishbein, M. and Ajzen, I.: *Belief, Attitude, Intention and Behavior: An Introduction to Theory and Research.* Addison-Wesley Publishing Company, Reading, Mass. 1975.

Fishbein, M. and Raven, B.: The AB scales: an operational definition of belief and attitude. In Fishbein, M. (Ed.): Readings in Attitude, Theory & Measurement. Wiley and Sons, New York, 1967.

Snider, J. G. and Osgood, C. E.: *Semantic Differential Technique. A Sourcebook.* Aldine, Chicago, 1972.

Stundner, G.: *Die Sicherheitseinstellung von Berufsschülern und deren Überprüfung durch tatsächliches Arbeitsverhalten.* Doctoral dissertation, Wein, 1977.

Weingarten, P.: Psychologische Aspekte Zur Unfallprophylaxe Bei Jugendlichen. Hefte Zur Unfallheilkunde, 1978.

Wicker, A. W.: Attitudes versus actions: the relationship of verbal and overt behavioral responses to attitude objects. *J Soc Issues, 25*, 41–78, 1969.

Zemanek, M.: *Motivationen zur Heimtierhaltung.* Doctoral dissertation, 1979.

SECTION II.

MEDICAL APPLICATIONS OF THE HUMAN–COMPANION ANIMAL BOND

CHAPTER 6

THE PET AS PROSTHESIS—DEFINING CRITERIA
FOR THE ADJUNCTIVE USE OF COMPANION ANIMALS
IN THE TREATMENT OF MEDICALLY ILL, DEPRESSED OUTPATIENTS*

MICHAEL J. McCULLOCH

There are at present no established criteria for the adjunctive use of pets in the treatment of medically ill depressed outpatients. No guidelines are available to help the clinician decide when a pet should be recommended or for what conditions or what type of pet to recommend.

Throughout history, companion animals have played an important role in the lives of individuals and families (Osler, 1976). But it has only been in the last quarter century, however, that serious scientific efforts have been made to describe the complex nature of the human-companion animal bond (Corson, 1975; Fox, 1975; Levinson, 1969b; Siegel, 1972). The observations of the practising veterinarian have added to the anecdotal literature.

THE TRIPARTITE BOND

The veterinarian, like all other health and allied professionals, is linked to companion animals and their owners through a complicated triangular relationship (McCulloch, 1977). (Fig. 6-1). Because of the unique opportunity to witness interactions of pets and pet-owning persons in families, the veterinarian has been placed in a role that has direct effect on human mental health (Antelyes, 1969a; Hopkins, 1978; McCulloch, 1979; Speck, 1964). Reactions of pet owners to the loss of pets have been explored (Keddie, 1977; Rynearson, 1978). The practising veterinarian is often placed in a position to observe such reactions, especially surrounding the

*The author wishes to thank Gregg Reiter, Ph.D., Shirlie Padgett, Jon Nash, and Sharon Goodrich for invaluable help.

101

decision to euthanize pet animals. In a study of seven veterinary practices, the number of euthanasias done to the total number of animal patients seen ranged from 1.8 to 3.6 percent, with an average of 2.1 percent. Thus, approximately one in every fifty interactions ends in a completed euthanasia (McCulloch, 1979). This does not include the number of times euthanasia of pets is discussed but not done.

FIGURE 6–1

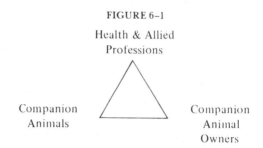

Additional use of animals in aiding deaf and blind persons has also been documented. The use of pets in a hospital for the criminally insane has been reported to improve the morale of staff members, reduce the incidence of inmate violence, and promote cooperative relationships among patients and staff (Lee, 1978).

The most thoroughly controlled study of the influence of pets on the emotional well-being of noninstitutionalized persons was done by Mugford and M'Comisky (1974). They gave elderly persons living alone either a budgerigar (small Australian parrot), a house plant, or nothing (control group). Extensive follow-up over a five month period revealed significant improvement in measures of self-esteem and emotional well-being in those given the budgerigars.

Although initial use of pets was primarily in the treatment of patients with psychiatric disorders, recent data indicates that pets may positively influence survival from certain medical illnesses such as myocardial infarction (Friedmann, 1979; Katcher, 1979). This fascinating finding has prompted considerable interest in the role that companion animals play in helping persons cope with medical illnesses.

The vast majority of people who are medically and/or emotionally ill live outside of institutional settings. Their care is provided through outpatient contact with doctors, other caregivers, and day use of institutional facilities. Many persons work and maintain relatively normal family lives in spite of illness. However, the range of disability is great; from partial and intermittent, to total and permanent. The causes of disability are varied and include acute and chronic medical conditions as well as traumatic injuries and their sequelae.

ADJUNCTIVE THERAPY

Most investigators who have used companion animals in the treatment of human conditions have emphasized that pet therapy is adjunctive only — not a substitute for other treatment methods. Little is understood yet about the effect pets have on the lives of persons who suffer from chronic medical illness. The frequency of depressive reactions accompanying physical illness is also not known but is thought to be significant.

As is often the case in clinical medicine, one occasionally discovers new insights by experimenting with various treatment methods whose criteria or efficacy may not be well documented but may offer relief from suffering. The following case illustrates the use of pets in the treatment of a depressed patient with a severe medical illness.

Mr. E. B. is a fifty-six-year-old married father of six who was referred for psychiatric evaluation because of depression. He was essentially in good health until he developed severe nephritis in 1972 at which time his renal function rapidly deteriorated. By 1976, he required renal dialysis. He received a kidney transplant from his son in 1976, but this was rejected and he was returned to dialysis. In May of 1977, a second transplant was attempted from a cadaver; but this was also rejected, and he once again returned to dialysis. He received a medical retirement in 1977.

In 1978, he suffered a myocardial infarction and continued to have angina on exertion from that time forward. By January of 1979, he was transferred from a hospital based outpatient dialysis unit to home dialysis. He was noted by his internist to be increasingly despondent and was placed on a low dose of antidepressant, which he tolerated fairly well in spite of his dialysis. However, his mood continued to deteriorate; he was noted to be very irritable with his family and very belligerent and argumentative. He was also observed to withdraw from other family members and friends. He did not enjoy his usual interests. His wife had gone to work which left him at home alone during the day.

Although he was collecting a disability income, he and his wife had completely reversed roles. She went off to work and obtained the paycheck, and he stayed home. He reported feeling increasingly useless, very angry at his physical restrictions, and imprisoned by his dialysis machine. He began to view himself as a burden to the rest of his family, and at times he wished that he would die while on the machine. His life was devoid of humor and everything seemed morbid and gray.

A review of his pet-owning experiences was made. As a child from age four through high school years, he and his sister had dogs. During his time in the military services he did not own pets because of travel and multiple moves. In 1965, he obtained a small basset hound that was killed after three months. The following year they obtained another dog that they had for four years. This animal was also killed. There had been no pets in the family since 1970.

When asked about his previous experience with pets, the patient replied that he greatly enjoyed pet companions as a child and also as an adult. It was suggested to him that he consider obtaining another pet dog to help distract him from his morbid preoccupation with his medical condition. After discussing it with his family, they agreed that it would be worthwhile to obtain another dog.

Considerable time was spent in finding the right animal as he wished to get another basset pup. He became very much interested in contacting various dog breeders and finally going to pick out the dog. Within two weeks, his spirits were improved, his activity level was increased, and with the arrival of the seven-week-old pup the tension in the household markedly decreased. The dog became a natural focus for family members. The antics of the animal caused laughter that had been conspicuously absent in the household for many months.

The patient's spirits continued to improve. He was noticeably less angry and seemed very involved and interested in the training of the animal. The patient's communication with other family members took on a much more positive note, and he reported feeling much less preoccupied with his illness and was more willing to be physically active in walking and training the dog. He also stated that it was nice to be needed again. The patient continued to do well on his dialysis but suffered intermittent angina and later had another myocardial infarction. He survived that and has continued on home dialysis every other day. He has continued his low dose of antidepressants, and he has remained absorbed in his pet dog which he has named "Hope."

The dramatic improvements were unforeseen in this case and prompted this speculation. Could the specific experiences of this patient be generalized to develop criteria for using pets to aid others suffering from medical illness and depression? To determine this, it was necessary to better understand how persons with chronic illness and depression perceive the impact of their illness on their lives, what they perceive their support system to be, and finally how they perceive the influence of pets on their lives and their illness.

If pets are to be prescribed for human ailments, they should be subject to the same scientific indications as are surgical procedures, drug therapy, and other forms of medical and psychiatric treatment.

METHODS

Medical records of fifty patients seen in a private psychiatric clinic during the past three years were reviewed. Criteria for inclusion in the study were the presence of a medical diagnosis and a depressive reaction. Patients were either seen in the hospital as a psychiatric consultation requested by another physician and later followed as an outpatient or referred directly for office consultation and treatment. To qualify for the study, the patients also had to have a pet in the household some of the time during which they were physically ill and depressed. Of the fifty patients, forty-five were available for interview. Of the forty-five patients contacted, fourteen did not have pets, leaving thirty-one who actually qualified for the study.

Questionnaires were given and included information about the nature and duration of the patient's physical illness and depression, the number and kind of pets, and the length of time pets had been in the household. This was followed by a fifty-three statement true/false questionnaire containing items on pet-owning history and experiences, the nature and depth of depressive symptoms, and the degree of incapacity secondary to physical illness.

Statements pertaining to the activity that promoted or hindered their rehabilitation were also included. Statements regarding positive and negative influences of pets were framed in an equal number so that the questionnaire was not biased to reflect a particular influence of pets.

There were fifteen men included in the study, ranging in age from twenty-eight to seventy-eight with an average age of fifty-two. Included in the study were sixteen women, ranging in age from twenty-two to sixty-nine with an average age of forty-six. Only six of the thirty-one were not married. Only one person was actually living alone. Many different medical illnesses were represented among the thirty-one patients. Many had multiple diagnoses (Figure 6-2).

FIGURE 6–2
NATURE OF ILLNESS

Cardiovascular
1. Myocardial infarction—coronary bypass
2. Myocardial infarction—coronary bypass
3. Myocardial infarction
4. Angina pectoris—coronary bypass—persistent angina

Cerebrovascular
1. CVA—mild memory deficit, left hemiparesis
2. CVA—left hemiparesis, ulcerative colitis
3. CVA—disruption of taste and smell

Traumatic Injury with Permanent Impairment
1. Knee fracture—dislocation
2. Shoulder and wrist injury
3. Compression fracture of spine
4. Degenerative arthritis of spine and knee
5. Total knee replacement, herniated lumbar disc, degenerative arthritis
6. Ruptured lumbar disc
7. Back sprain

Endocrine
1. Diabetes
2. Diabetes, hypertension
3. Diabetes, hypothyroidism
4. Cushing's syndrome
5. Hypoestrogenism

Gastrointestinal
1. Peptic ulcer disease
2. Peptic ulcer disease with multiple surgeries
3. Intestinal bypass surgery, dumping syndrome, back and neck sprain
4. Multiple abdominal surgeries, temporal arteritis, compression fractures of spine

Miscellaneous
1. Lymphatic cancer
2. Giant urticaria
3. Chronic recurrent pneumonia
4. Parkinson's syndrome
5. Renal failure—dialysis
6. Shoulder-hand syndrome, arthritis
7. Trigeminal neuralgia
8. Emphysema

There were four patients who had heart disease; three of these had experienced coronary bypass surgery, and their condition had noticeably improved since surgery. There were three persons who had suffered cerebral vascular accidents with permanent impairment from mild to moderate.

Of the thirty-one patients, seven suffered traumatic injury with permanent impairment. Several of these were work related injuries, others were motor vehicle accidents. All of these conditions had stabilized at their present level of disability, and each of these injuries was accompanied by chronic pain.

There were five persons who had endocrine disorders and were among the least incapacitated of all thirty-one patients due to good control of their condition with medication or proper diet.

There were four people with gastrointestinal disorders—with two having chronic ulcer conditions. One patient had a persistent dumping syndrome from intestinal bypass surgery for obesity. A fourth patient experienced multiple abdominal surgeries and complications in addition to temporal arteritis and compression fractures of the spine secondary to steroid therapy.

A miscellaneous group of illnesses included lymphatic cancer and giant urticaria, a severe debilitating allergic condition. One patient had end-stage renal disease and was on dialysis. Another suffered from a rare painful condition known as shoulder-hand syndrome with progressive loss of function accompanied by arthritis. And one patient had very severe emphysema with chronic recurring respiratory infections.

There was considerable variation in the severity and duration of disability. The heart problems caused severe disability for a short period of time and some limitations after that. Those with traumatic injuries exhibited the most pronounced loss of function over time. The endocrine conditions were the least disabling. The remaining conditions varied greatly in the extent of disability.

Figure 6-3 illustrates the number of pets in each household. There were seventeen patients who had one or more dogs; five patients had one or more cats; six patients had a combination of dogs and cats; and three patients had, in addition to dogs and cats, birds, hamsters, and other rodents. No exotic pets were reported.

FIGURE 6–3
PETS IN HOUSEHOLD

Dogs	17
Cats	5
Dog(s) and Cat(s)	6
Dogs, Cats, Birds, etc.	3

The effects of illness on individual patients is described in Figure 6-4. Of the thirty-one patients, twenty-seven reported that they did not enjoy their usual interests, nor could they be as physically active as before their illness. There were twenty-five patients who indicated that they were unable to work some of the time and felt like a burden to other people. There were twenty patients who said they spent a lot of time alone and also worried that their condition would not improve. There were eighteen patients who reported being fearful, and fifteen patients who reported being totally unable to work and were financially dependent on others.

FIGURE 6–4
EFFECTS OF ILLNESS—31 PATIENTS

1. Totally unable to work	15 patients
2. Unable to work some of the time	25 patients
3. Could not be as physically active	27 patients
4. Felt like burden to others	25 patients
5. Financially dependant on others	15 patients
6. Spent a lot of time alone	20 patients
7. Fearful	18 patients
8. Did not enjoy usual interests	27 patients
9. Worried that condition would not improve	20 patients

The patients were also questioned as to the factors they felt were the most supportive in helping them cope with their illness. These are outlined in Figure 6-5.

FIGURE 6–5
SUPPORT FACTORS DURING ILLNESS – 31 PATIENTS

1. Spouse or close personal relationship	26 patients
2. Resuming work duties	26 patients
3. Family	25 patients
4. Pet was an important source of companionship	26 patients
5. Pet improved my morale	20 patients
6. Hobbies, athletic interests	20 patients
7. Contact with friends	19 patients
8. Religious faith	15 patients

There were twenty-six patients who reported a spouse or close personal relationship and resuming work duties as important factors. There were twenty-five patients who reported that family support was helpful. There were twenty who indicated that hobbies and athletic interests were important. There were twenty who reported that a pet helped improve their morale. There were nineteen patients who indicated that contact with friends was welcomed, while fifteen patients indicated their faith was of value to them. There were twenty-six patients who also listed "pet" as an important source of companionship during illness.

An interesting aspect to this study was a fact that only fifteen patients of the thirty-one reported having the primary bond with the pet. There were sixteen patients who reported that other family members had the primary emotional relationship with the pet. These included pets that had belonged to children who had moved out of the home as well as pets that belonged to other members of the family. In spite of the reported differences in emotional attachment, both groups had a high positive response to several statements regarding pets being an important part of their lives, a pet's affection being greatly appreciated, pets helping cope with isolation and loneliness, and pets helping them to laugh and maintain a sense of humor. Encouraging a sense of humor was the highest reported influence by both groups. (Fig. 6-6)

FIGURE 6-6

INFLUENCE OF PETS	PATIENT HAS PRIMARY BOND WITH PET 15 PATIENTS	OTHERS HAVE PRIMARY BOND WITH PET 16 PATIENTS	TOTAL 31 PATIENTS
1. I am my pet's closest companion	15	16	31
2. Pets have been an important part of my life	14	13	27
3. Pet an important source of companionship during illness	13	9	22
4. Pet made me feel needed	12	9	21
5. Pet distracted me from worry about problems	12	8	20
6. Felt more secure with pet around	12	10	22
7. Pet improved my morale	11	9	20
8. Pet's affection greatly appreciated	15	14	29
9. Pet's playfulness improved my spirits	12	11	23
10. Pet helped me cope with isolation and loneliness	13	13	26
11. Pet helped me to laugh and maintain a sense of humor	15	14	29
12. Pet's needs stimulated me to be more physically active	15	6	21

There were differences between the two groups. Patients with the primary emotional bond with their pets reported pets as being an important source of companionship during illness; pets made them feel needed and more secure and also helped distract them from their problems. Pets were also credited with improving morale as well as providing stimulation to be physically more active.

A number of negative factors regarding the influence of pets in the household were obtained in the questionnaire and are outlined in Figure 6-7. Very few reported having mostly bad experiences with pet ownership. There were four who reported that the pet was too much responsibility when they were ill. Only two wished that pets had not been in the household. There were seven patients who reported that the pet at times was a nuisance, and eight worried about the pet's care if their illness became worse. It is significant that sixteen patients in this study reported that at times they were too ill to care whether a pet was around. The responsibility most likely fell to others during that time.

FIGURE 6-7
NEGATIVE FACTORS—31 PATIENTS

1. Too ill at times to care whether pet was around	16 patients
2. Worried how pet would be cared for if illness became worse	8 patients
3. Pet was a nuisance	7 patients
4. Pet was too much responsibility	4 patients
5. Wish pet(s) had not been in household during illness	2 patients
6. Got angry at pet	1 patient
7. Have had mostly bad experiences with pet ownership in the past	1 patient

DISCUSSION

There are a number of interesting findings in this study. Most of the thirty-one patients had adequate support systems in the form of family or close friends. Many were also able to return to work on at least a part-time basis, and for some their disabilities greatly improved to a point where they could resume a nearly work on at least a part-time basis, and for some their disabilities greatly improved to a point where they could resume a nearly normal life-style. Only one of the thirty-one patients was divorced at the time of the study; two had lost their spouse through death. Only one patient among the thirty-one was actually living alone, and she was widowed.

In spite of the relative social stability of the sample and the seeming adequacy of support systems, the presence of pets was perceived as an important addition toward coping with illness and depression. If the patient were living alone, had a greater disability and less intact support system, then one could expect the pet to assume even greater importance.

Among the diseases represented in the study, there were wide variations in the severity and duration of disability. Those suffering from cardiovascular disease fortunately recovered from their myocardial infarction, and three of the four patients underwent a successful coronary bypass surgery with marked relief of symptoms. None of the patients in this group suffered prolonged problems with congestive failure, arrythmias, or other serious complications. This group cannot be considered typical of those who struggle with problems of chronic heart disease, as the degree of disability was relatively low. However, depression was prominent.

There were other illnesses that had permanent levels of impairment. Those included three patients with a stroke. All three patients were ambulatory and functional but did experience feelings of loss and prolonged problems with depression. Of the three patients, one had attempted suicide.

The largest single category of disability included patients who had experienced traumatic injury, all with permanent impairment. Level of disability in each patient had led to loss of work time. Of the seven patients, four had to completely change occupations and be retrained. Work related injuries are a significant cause of disability, and only at a much later date is depression often recognized and appropriate referrals made for treatment.

Endocrine disorders as a group were not as disabling and were generally controllable medically. There was one patient who suffered from severe Cushing's syndrome and required bilateral adrenalectomy and cortisone maintenance thereafter.

Gastrointestinal disorders were more disabling in this study than the endocrine problems and accounted for a greater number of hospitalizations and time loss from work. This group also accounted for the greater number of surgeries per patient.

Among the miscellaneous diseases, giant urticaria, an unusual allergic condition with persistent giant hives, was extremely disabling and led to severe depression. The patient with Parkinson's syndrome was fairly well controlled with medication, but the medication itself tended to promote depression and loss of function.

There was one person in the study who had end-stage renal disease at an early age and had experienced a transplant failure. Dialysis was required every other day. This patient reported that she was frequently sick and worried about her own future. Her interest in her pets was extremely variable.

The patient with shoulder-hand syndrome, a rare but severely disabling condition of the upper extremities accompanied by pain and loss of function, had to euthanize her two dogs. During the year afterward, her depression worsened considerably, and she was alone most of the day. Her mood improved when two new dogs were obtained for her by her husband.

The patient with emphysema was an elderly widower living with his son. He spent most of his time alone during the day. Following the death of his wife, he became extremely dependent upon his pet dog who was his constant companion. He frequently acknowledged to family members that if it were not for the dog, he would not have the will to go on.

There are many severely disabling illnesses that are not represented in this sample and require further study. These include various forms of cancer that often require prolonged chemotherapy and/or radiation therapy. They are severely disabling, very often progressive, and frequently accompanied by severe depression.

All of the subjects in this study experienced marked life change with their illness and subsequent disability. The literature on the effects of life change on the incidence and severity of illness is well documented (Holmes, 1967). Recent evidence also suggests that certain character traits may make a person less vulnerable to illness during life change (Kobasa et al., 1979). Nonetheless, life change does cause a period of vulnerability and among many of

the subjects appeared to increase their dependence on pets as well
as other elements of their support systems. Because of this increased
dependence, it is strongly suspected that such persons would be
more vulnerable to loss of their companion animals.

The effects of illness were very apparent in the lives of these
thirty-one patients. The vast majority reported that they could
not be as physically active and that their illness had caused them
to be unable to work some of the time. Over half of the patients
stated that they could not work at all during their disabling
illness. There were two-thirds of the patients who spent a considerable amount of time alone, and at least half of the sample
indicated that they had become financially dependent upon others
because of their disability. Loss of interest in activities that were
previously enjoyable was also reported by the majority of patients.

Analysis of the reported support factors during illness revealed
that support of spouse or close personal relationships and resuming work duties ranked as the most important factors that helped
them cope with their illness. Of interest is that pets were reported
to be an important source of companionship. No effort was made
to discriminate the relative degree of importance that each person
placed on the various support factors. That question remains for
future study. Family, contact with friends, hobbies, athletic interests, and religious faith emerged as the other significant elements
of support.

The presence of pets in the household during illness of the
study subjects evoked relatively few negative responses. For example, four felt that having pets in the household was too much
responsibility, while only two actually wished there had been no
pets in the household at all. Pets were considered a nuisance at
times by seven patients.

Perhaps the most significant negative factor was that sixteen
patients reported they were too ill at times to care whether a
pet was present in a household. This suggests that an adequate
support system for caring for pets needs to be present, especially
in those persons whose illness is likely to deregulate and cause
serious disability or loss of function. This would definitely include
rehospitalization, which for many pet owners creates difficulty,
requiring that the pet be maintained by other family members or
boarded.

The findings of the study appear to confirm previous studies that describe the psychological benefits of pet ownership (Mugford, 1979a) (Fig. 6-8). In his thorough review of the literature, companionship is reported to be the most significant reason for acquiring and maintaining pets. There are two measures of companionship—affiliation and self-esteem—that emerge as significant determinants of pet ownership. Affiliation refers to the desire for communication, friendly interaction, and close physical proximity to other living things. Among the patients in this survey reporting primary relationships with pets, this finding is reaffirmed. Self-esteem, although a complex concept, in general refers to contentment with one's self and being appreciated, wanted, or loved by others. In this study, twenty of the thirty-one patients reported that the presence of pets improved their morale, and twenty-one reported that pets made them feel needed.

FIGURE 6–8

PSYCHOLOGICAL BENEFITS

Companionship
    Affiliation
    Self-Esteem

Play
Attachment and Love
Emotional Security
Child Substitute

Play emerges in the literature as a very significant contribution of contact with pets, and its influence on human health has been vastly underrated. There were twenty-three patients who reported that their pet's playfulness improved their spirits. There were twenty-nine who indicated that a pet helped them to laugh and maintain a sense of humor. It is strongly suspected that the element of play is directly related to promoting a sense of humor.

Other writers have spoken eloquently about the role of humor in coping with illness. In his book, *Anatomy of an Illness as Perceived by the Patient—Reflections on Healing and Regeneration*, Norman Cousins described the role of laughter in helping him to reduce the amount of pain that he experienced. He further went on to say that his sedimentation rate, a measure of the severe inflammatory process that affected him, was actually reduced by his systematic program of introducing humor into his rehabilitation.

He further stated, "I was greatly elated by the discovery that there is a physiologic basis for the ancient theory that laughter is good medicine". The only negative effect reported by Mr. Cousins was that his laughter program was disturbing the other patients in the hospital.

One of the considerations for further study is the examination of pets' role in promoting humor in those suffering from illness. If humor is determined to have positive influence on illness and depression, then this may be one of the pet's most significant contributions to human health.

Among the other psychological benefits are attachment and love. In this sample, twenty-nine out of thirty-one reported that the pet's affection was greatly appreciated. Because a sick person's self-esteem is frequently adversely affected, having a pet depend on them for basic needs is one way that persons can cope with their own negative feelings about dependency. The nature of the physical contact with pets may also have a physiologically meas-urable calming effect. Whether this is caused by distraction from worry about problems, as twenty patients in the study indicated, or through some other avenue is not clear.

Emotional security is another reported psychological benefit of having pets. For persons suffering with chronic illness, there is often the fear or dislike of being alone. Of the thirty-one patients, twenty-two reported that they felt more secure having a pet in the household. It is likely that the nature of the emotional bond between persons and their pets makes them feel more secure in their physical surroundings as relatively few animals are actually able to provide true physical security. Child substitute reported in the literature as a psychological benefit of maintaining pets was not examined in this study.

The need for companionship, especially during chronic illness and depression, has been noted by Lynch in his book, *The Broken Heart, The Medical Consequences of Loneliness*. In this work he states that, "The lack of human companionship, the sudden loss of love, and chronic human loneliness are significant contributors to serious disease (including cardiovascular disease) and premature death." He alludes to the importance of the quality of interaction, not just the presence of another human being in the life of someone who is ill as being an essential ingredient. He describes the need for "dialogue" between human beings as a critical element in maintaining physical and emotional well-being. However, he stops short of considering whether the quality of interaction with animals can have an ameliorating effect on disease and emotional health.

Friedman and Rosenman, in their book, *Type A Behavior and your Heart*, also consider that the increase of human loneliness may be a prominent source of cardiovascular problems. They reaffirm that just having another person physically present, such as in a dysfunctional marriage, does not prevent an individual from experiencing acute loneliness and social isolation. Whether animals can have an ameliorating effect on dysfunctional relationships and family difficulties requires further study but is strongly suggested by the clinical case discussed in this paper. Pets may indeed have a neutralizing effect on conflict through their ability to distract, entertain through play, and promote laughter.

Depression is a serious complication of many physical illnesses. Among the elderly, physical deterioration and losses increase the likelihood of depression. Suicide rates among those approaching the elderly years are alarmingly high. The influence of loneliness, isolation, and fear has been eloquently described (Butler, 1975). For some physical illnesses, depression is a predictable consequence. In a study by Stern et al., 17.3 percent of patients suffering from a myocardial infarction were depressed at six weeks. Without treatment, 15.9 percent remained depressed at one year. The impact of depression as a complication to medical illness remains one of the challenging treatment dilemmas for the health and allied professions.

CONCLUSION

Review of data in this study strongly suggests that the subjects experienced feelings about their illness and disability similar to the patient in the case study (Fig. 6-4). The study group of thirty-one patients reported several important support factors during illness (Fig. 6-5) that were different from those cited by our case subject. He felt alienated from virtually all aspects of his support system with the notable exception of his pet, which acted as a catalyst to reopen access to this system. The case subject and the study group shared similar perceptions of the influence of their pets on the illness and depression (Fig. 6-6).

There are many clinical indications for which prescribing pets may be very appropriate. The following is an attempt to outline one set of specific indications. It is hoped that such criteria will provide some guidance to clinicians who are confronted each day with complex treatment dilemmas of patients with chronic illness.

It must be reemphasized that the use of pets is recommended in an adjunctive way and is not aimed at displacing other legitimate forms of therapy. It is intended to complement them — to be a prosthesis.

1. Chronic illness or disability

2. Depression

3. Positive previous relationship with pets

4. Role reversal

5. Negative dependency

6. Loneliness, isolation

7. Helplessness

8. Low self-esteem

9. Hopelessness

10. Absence of humor

These criteria can be seen in many different illnesses. Virtually any disease that causes a person to be off work, and/or in a financially dependent position will affect self-image, promote feelings of helplessness, hopelessness, and depression. If it is accompanied by a role reversal with spouse or significant person, this may further promote a negative or angry dependency. Loss of sense of humour frequently accompanies depression as well. If the person reports a positive past relationship with pets, then a similar pet can be added. If such a patient already has a pet, the value of the pet can be reemphasized so that the person becomes more consciously aware of the potential benefits.

Suggested Precautions

1.   *Beware of increased vulnerability to the loss of a pet.* Adding a pet to a depressed individual's life can be beneficial, but one must appreciate that in that state of mind there is also an increased state of vulnerability to loss of pet. For such a person, a superimposed loss of this type can be hazardous.

2.   *Tailor the prescription pet to the individual.* No attempt has been made in this study to discuss pet matching. Other writers have begun to outline more specific methods of appropriately matching pets to perspective pet owners (Bustad, 1979). Further investigation is required in this area to ensure that guidelines for responsible pet ownership are promoted and followed.

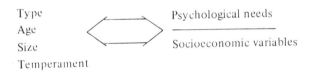

3.   *Coordinate the use of prescription pets with other therapy methods.* Simply adding a pet to a person's life is not sufficient. One must coordinate its introduction with other therapy approaches, including medication, family intervention, or individual therapy. For many persons, physical therapy, as well as other highly tailored rehabilitation programs, needs to continue. It is hoped a pet would be a catalyst in many of these situations and act as a tension breaker within the family unit.

4.   *Identify situations that are inappropriate for prescription pets.* For many persons their physical or emotional condition does not permit adequately incorporating a pet into their life. Persons who suffer from organic brain impairment with loss of memory, confusion, and poor judgment should not have full responsibility for

pet care. If supervision cannot be adequately arranged, then some form of part-time exposure to pets should be attempted. There are instances in which a person may be too depressed to care for a pet. Level of function of the depressed person must be assessed before a pet is added.

5. *Timing.* As with most therapies, timing is essential. A pet should be given at an appropriate time when it can be incorporated and its benefits can be most appreciated. This strongly suggests that a person's physical condition be relatively stable so that the contact with the pet can be fairly consistent. If the person's condition is extremely unstable, one must be certain that the pet can be properly cared for in a consistent fashion while the person is being treated.

Many questions arise from a study of this type. Virtually every one of the illnesses or disabilities could warrant separate study and more extensive examination. One area that has not been adequately discussed is the role that pets might play in the lives of sick, developmentally disabled, or injured children. Children also suffer terminal illnesses, kidney failure, diabetes, severe injuries, and other problems. However, their personality development continues through the time of disability. What role pets might play in helping to maintain a fairly normal psychosocial development in such disabled children remains a subject for future investigation.

We must, as health and allied professionals, be aware of the continuous struggle to maintain self-esteem and dignity in the face of chronic debilitating illness. Many patients require a careful reorientation to their disability. For most it is hard to forget "the way I used to be." They mourn their loss of function and often set unrealistic expectations of their improvement only to be disappointed and lapse into despair. We must be prepared to help them use the lowest point of their disability as their new frame of reference. Then all progress will seem to be a relative success. If they compare their progress with the way they used to function, they will continually fall short and experience failure.

Rehabilitation and readaption is a struggle of inches and feet, not yards and miles. We must continue our efforts to promote a humane treatment environment, one in which comforting and caring are essential ingredients.

We are often inclined to look at the unique characteristics of disease entities rather than the common features of disabling illness. We must be aware that "the meaning of being disabled includes an awesome recognition that organs and body parts are no longer conspicuous compliant instruments that carry out their owner's intentions.... the sick person becomes an object to be manipulated by forces beyond comprehension and control" (Weisman, 1974).

In his paper, "Chronic Disabling Illness: A Holistic View," Feldman states "regardless of previous life experience, to be the victim of catastrophic illness poses an awesome reality. The stricken one must accept that life is irrevocably different. Because of that difference, a new meaning and way of life must be found.... to discover a new meaning in life in the face of the dissolution of the old meaning, to accept a difference imposed by the illness, and to still maintain one's dignity and worth is the essence of the transition of sick to [being] different." If we truly believe Rudolph Virchow's statement that "medicine is a social science in its bone and marrow," those of us in the health professions must continue to help patients maintain optimism and the will to survive.

In an age of research when it is tempting to reduce emotions to biochemical reactions and to rely heavily on the technology of medicine, it is refreshing to find that a person's health and well-being may be improved by prescribing contact with other living things. Members of the health and allied professions must continue to combine resources, work together in the spirit of cooperation, and never forget to "cure when possible but comfort always."

REFERENCES

Antelyes, J.: The troubled pet-owner visits the veterinarian. *Veterinary Medicine/Small Animal Clinician*, 38–40, January 1969a.

———: The perfect pet-owner and other fantasies. *VM/SAC*, 315–18, April 1969b.

Bustad, L. K.: Profiling animals for therapy. *Western Veterinarian*, 17 (1): 2, 1979.

Butler, R. N.: *Why Survive? Being Old in America.* Harper and Row, New York, 1975.

Carithers, C. M.: Pets in the home: incidence and significance. *Pediatrics*, May 1958.

———: Children and their pets. *JAMA, 14*, no. 5, May 1959.

Corson, S. A. et al.: Pet-facilitated psychotherapy. In Anderson, R. S. (Ed.): *Pet Animals and Society.* The Williams and Wilkins Co., Baltimore, 1975, pp. 19–36.

_____ : The socializing role of pet animals in nursing homes: an experiment in nonverbal communication therapy. *Proceedings of the International Symposium on Society, Stress and Disease: Aging and Old Age.* June 14–19, 1976.

_____ : Pet dogs as nonverbal communication links in hospital psychiatry. *Comprehensive Psychiatry, 18*: 61–72, 1977.

_____ : Pets as mediators of therapy in custodial institutions and the aged. In Masserman, J. H. (Ed.): *Current Psychiatric Therapies,* vol. 18. Grune and Stratton, New York. 1979.

Cousins, Norman: *Anatomy of an Illness as Perceived by the Patient: Reflections on Healing and Regeneration.* W. W. Norton and Co., New York, 1979.

Feldman, D. J.: Chronic disabling illness: a holistic view. *J Chron Dis, 27*: 287–91, 1974.

Fox, M. W.: Pet-owner relations. In Anderson, R. S. (Ed.). *Pet Animals and Society.* Bailliere Tindall, London, 1975, p. 37.

Friedman, M. and Rosenman, R.: *Type A Behaviour and Your Heart,* Alfred A. Knopf, New York, 1974.

Friedmann, E., Pet ownership and survival after coronary heart disease. *Presented at the 2nd Canadian Symposium on Pets and Society.* Vancouver, B.C., Canada, May 30–June 1, 1979.

Heiman, M.: The Relationship Between Man and Dog. *Psychoanal Quart, 25*: 568–85, 1956.

_____ : Man and his pet. In Slovenko, Ralph and Knight, James A. (Eds.): *Motivation in Play, Games and Sports,* Thomas, Springfield, 1967.

Holmes, T. H. and Rahe, R. H.: The social readjustment rating scale. *J Psychosom Res, 11*: 213–18, 1967.

Hopkins, A. F.: Ethical implications in issues and decisions in companion animal medicine. In McCullough, L. B. and Morris, J. P. III (eds.): *Implications of History and Ethics to Medicine – Veterinary and Human.* Centennial Academic Assembly, Texas A and M University, College Station, Texas, 1978.

Katcher, A. H.: Social support and health: effects of pet ownership. *Proceedings of the Meeting of Group for the Study of Human Companion Animal Bond.* Dundee, England, March 23–25, 1979.

Kobasa, S. C., Hilkes, R. J., and Maddi, S. R.: Who stays healthy under stress? *J Occupational Medicine, 21*(9): 595–98, September 1979.

Keddie, Kenneth M. G.: Pathological mourning after the death of a domestic pet. *British Journal of Psychiatry, 131*: 21–25, 1977.

Lee, D.: *Hi Ya, Beautiful.* A documentary film on pet therapy at Lima State Hospital, The Latham Foundation, 1978.

Leigh, D.: The Psychology of the Pet Owner. *Journal of Small Animal Practice, 7*: 517–21, 1966.

Levinson, B. B.: The veterinarian and mental hygiene. *Mental Hygiene, 49*, July 1965.

_____ : Interpersonal relationships between pet and human being. In Fox, M. W.: *Abnormal Behaviour in Animals.* W. B. Saunders, Co., Philadelphia, 1968.

_____ : Pets and old age. *Ment Hyg, 53* (3), 1969a.

_____: *Pet Oriented Psychotherapy.* Thomas, Springfield, 1969b.

_____: Pets, child development & mental illness. *JAVMA, 157:* 1759–66, December 1, 1970.

_____: Man, animal, nature. *Modern Veterinary Practice, 53,* April 1972.

_____: *Pets and Human Development.* Thomas, Springfield, 1972.

_____: Pets and environment. In Anderson, R. S. (Ed.): *Pet Animals in Society.* Williams and Wilkins, Baltimore, 1974.

Lynch, James J.: *The Broken Heart: The Medical Consequences of Loneliness.* Basic Books, New York, 1977.

McCulloch, M. J.: *The Pet-Owner Relationship — Dealing with the Difficult Client.* Paper to American Veterinary Medical Association annual meeting, Anaheim, California, July 1975.

_____: Contribution of veterinarians to mental health. In *A Description of the Responsibilities of Veterinarians as They Relate Directly to Human Health.* U.S. Govt. Department of Health, Education & Welfare, Bureau of Health Manpower. Contract No. 231-76-0202, 1976.

_____: The veterinarian in the human health care system: issues and boundaries. In McCullough, L. B. and Morris, J. P. III (Eds.): *Implications of History and Ethics of Medicine — Veterinary and Human.* Centennial Academic Assembly. Texas A & M University College, Station, Texas, 1978.

_____: *Pets and Family Health.* Paper to the American Veterinary Medical Association Annual Meeting, Seattle, Washington, July 1979.

Mugford, R. S. and M'Comisky, E. G.: Some recent work on the psycho-therapeutic value of cage birds with old people. In Anderson, R. S. (Ed.): *Pet Animals and Society.* Bailliere Tindall, London, 1975.

Mugford, R. Basis of the normal and abnormal pet/owner bond. *Proceedings of Meeting of Group for the Study of Human Companion Animal Bond.* Dundee, Scotland, March 23–25, 1979a.

_____: The social significance of pet ownership. In Corson, S. A. (Ed.): *Ethology and Non-verbal Communication in Mental Health.* Pergamon Press, London, 1979b.

Osler, W.: The relations of animals to man. Lecture given at Montreal Veterinary College. *Vet J and Annals of Comp Path, 3:* 465–66, 1876.

Rynearson, E. K.: Humans and pets and attachment. *Br J Psychiatry, 133:* 550–55, 1978.

Siegal, A.: Reaching the severely withdrawn through pet therapy. *Am Journ of Psych, 118:* 1045–46, March 1962.

Speck, R. V.: Mental health problems involving the family, the pet, and the veterinarian. *JAVMA, 145* (2), July 1964.

Stern, M. J. et al.: Life adjustment postmyocardial infarction — determining predictive variables. *Arch Int Med, 137:* 1680–85, 1977.

Virchow, R.: quoted in "Chronic disabling illness: a holistic view, *J Chron Dis, 27:* 287–91, 1974.

Weisman, A.: quoted in Chronic disabling illness: a holistic view. *J Chron Dis, 27:* 287–91, 1974.

Wilbur, R. H.: *Pet Ownership and Animal Control, Social and Psychological Attitudes,* 1975. Report to the National Conference on Dog and Cat Control, Denver, Colorado, February 4, 1976.

CHAPTER 7

PET FACILITATED THERAPY IN HUMAN HEALTH CARE

JULES CASS

*P*et *facilitated therapy* (PFT) is the introduction of a pet animal into the immediate surroundings of an individual or a group as a medium for interaction and relationships, with the therapeutic purpose of eliciting physical, psychosocial, and emotional interactions and responses that are remedial. Other terms used for pet facilitated therapy are *pet mediated therapy*, *pet facilitated psychotherapy*, and *pet oriented psychotherapy*. Pets are described in their role as mediators in therapy as reinforcers, socializing catalysts, aids to therapy, cotherapists, pet companions, patient ward mascots, and psychological support systems. Pets can serve as a shortcut on the long road of therapy and as a medium for transference to human beings. Pets have been observed to serve in building a communication chain, linking the patient to the pet, then joining the patient to the professional therapist. For patients with certain conditions, pets can serve as an adjunct to other forms of therapy.

WHAT FORMS OF PFT ARE BEING PRACTICED TODAY?

PFT is being prescribed and practiced in a variety of settings, both noninstitutionalized and institutionalized. Outside of institutions there are physically disabled (the blind, the deaf, those confined to the wheelchair), and there are people at risk of becoming institutionalized (the physically or psychosocially impared, mentally disturbed or handicapped, the elderly, and even students in school dormitories). Inside of institutions (hospitals, nursing homes, homes for the elderly, prisons, schools for the retarded, facilities for the chronically ill and dying) there are the physi-

124

cally, psychosocially and mentally disturbed, the handicapped, the elderly, and the chronically ill and dying.

DOES PFT REALLY WORK AS A FORM OF TREATMENT?

Because the vast body of evidence in support of PFT is derived from clinical and anecdotal experiences and observations, there are those in human health care services who consider the practice of PFT as being interesting but at this point requiring added supportive evidence through research before reaching acceptance and clinical application.

Research, as defined by today's attitudes of acceptance of what is and what is not research, insists that it is not enough to make valid observations and to record them but that there must be a quantitation to assure that the observations did not occur by mere chance. The observations and descriptions of the experiences recorded in anecdotal writings currently constitute the major portion of evidence that PFT does work successfully in the areas reported.

The recording of single observations in biology was a foundation block of knowledge in all biological and medical sciences. Jean Henri Fabre, naturalist-entomologist, left behind a rich and essential legacy of observations made as he sat for hours capturing and recording the movements, interactions, and life's activities of social and solitary bees and wasps. His understanding of the relationship of each of these observations to the whole was not always complete, but the single observation in itself was valid.

In the use of drugs, man and animals are correctly and essentially protected against drug use clinically until a drug is certified through research as biologically safe and effective (at the dosages prescribed) and with insignificant side effects.

The drug lithium chloride was approved as a safe and effective drug for use in successfully ameliorating human maniacal depression but had not been successful during clinical trials and, therefore, not recommended for the treatment of other forms of depression. Yet, clinical psychiatrists started to and are now prescribing this drug, within a safe dosage range, to some of their depressed patients (not maniacal) with therapeutic successes. Why? No one has the foggiest notion. Of course, clinical studies and observations continue to be carried out, along with fundamental pharmacologic research on lithium and its mechanism of action. However, lack of this research data did not slow down the increasing use of this drug for therapy of depressed conditions in some patients, for which no rationale of effectiveness had been identified in the original biological evaluation. It is the clinical results benefiting some patients that warrant use of this approved drug. The verification and sharpening-up of its use through research will continue to follow. But therapists are not waiting for the definitive research results. Selected patients have benefited and have apparently not been harmed. The accumulation of successful clinical experience and observations becomes a valid rationale for treatment until the supporting research data is produced.

Now I would like to draw a parallel example, applying the same reasoning to the empirical use of PFT as is being applied in the use of lithium. PFT, as a form of treatment, meets the requirements of certification as being biologically safe and effective (provided the essential elements of PFT are properly established and administered) and with no significant side effects. People with several defined physical, psychosocial, and mental health conditions are being administered PFT as a copreventive and as a form of cotreatment. There is a limited understanding of the conditions for interaction and the mechanisms of the interactional response that produce the beneficial results. Of course, clinical studies and observations continue to be carried out, along with some research on PFT, the latter to disclose the details for understanding and for more precise clinical use. Lack of this research data has not slowed, nor should it slow, the increasing use of PFT health care in some patients and institutional residents. Some of these special people are being helped and apparently not harmed. The accumulating experience and observations together with some limited research data become a rationale for the empirical use of PFT as a valid treatment form in human health care.

Nevertheless, every effort must be made for research investigators to initiate and continue to conduct systematic PFT research. Professional therapists, responsible for the physical and mental health care of individuals, have every right to ask for research evidence that supports and validates the current, empirical practices and uses of PFT. Such essential research data would assist them in understanding the underlying psychophysiologic bases for the observed changes in health, thereby enabling them to determine what practices and conditions of PFT would best be prescribed and when to prescribe them in planning the therapeutic strategy for a person or a group. Research will extend the data needed to answer questions like those raised as the objectives in the research program reported by Corson, O'Leary Corson, and Gwynne in 1975.

1. Is PFT feasible in a hospital setting?

2. Is it possible to develop procedures for matching different types of dogs with the personality and specific psychopathology of a particular patient?

3. What breeds are likely to be most suitable for different diagnostic categories?

4. Is it possible to develop practical training programs for the interaction between a dog with a given temperament and a particular patient?

5. Can PFT reduce length of hospital stay of patients?

6. How much saving in staff time may result from PFT?

7. Can the application of PFT reduce the frequency of restraining procedures on patients?

8. Can PFT reduce the frequency of forcible use of drugs?

9. Can PFT decrease the incidence of suicidal attempts or gestures in an inpatient population?

10. To what extent can PFT decrease dosage requirements and duration of drug administration?

11. To what extent can PFT facilitate meaningful therapeutic communication with patients?

12. Can we utilize videotape recordings of pet-patient interaction as a form of feedback therapy for patients?

13. Can we develop PFT training programs for residents and medical students including the use of videotape techniques?

14. Is it possible to develop diagnostic procedures based on the types of interaction between a patient and different types of dogs?

15. Is it possible to develop a training program of patient-pet interaction so that a patient can leave the hospital with a pet as a form of continuing outpatient therapy?

Some of these questions have begun to be answered by limited research documentation (Corson, 1975; 1975a; 1976; Brickel, 1979). Research should provide the data that will allow predicting the likelihood and levels of successful therapy that can be achieved with PFT alone or in combination with other traditional forms of therapy and the impact of programs of PFT on rehabilitation.

ENVIRONMENTAL SETTINGS FOR PFT

For the institutionalized resident or patient, the surrounding environment is an essential admixture of multiple factors. They constitute the physical and psychosocial milieu. These factors, living and inanimate, can affect the fabric and quality of life in unrecognized and unmeasured ways. Totally-cared-for chronically ill or disabled patients reside in institutional surroundings that are necessarily determined and maintained *in toto* by the physical design of the institution, by the administration and its operational practices in patient care, the therapist staff, and the psychosocial and physical quality of the surrounding factors in direct patient contact as well as those within range of the senses of seeing, hearing, touch, and smell. Such patients and residents are as totally dependent as are pet animals maintained indoors at all times.

An emphasis on integrating the environmental matrix (physical, psychosocial, emotional) of the individual patient into his treatment strategy has become a part of the Veterans Administration's policy. It has resulted in buildings particularly designed to incorporate humane facilities for the chronic care of elderly patients. Humanization of chronic care institutions, including psychiatric facilities (and hospices) is congruent with the stated goals of the U.S. Joint Commission for Accreditation of Hospitals. A sincere concern for the need to deinstitutionalize the new milieu by means of improving the psychosocial environment in patient care facilities has encouraged the development of detailed, comprehensive recommendations. For example, a current recommendation by the Veterans Administration Central Office Committee on the Quality of Life provides for an environment that allows patient access (and "can keep privileges") to *appropriate* pets *on wards* (Enquist, 1978; *Inpatient Program Guide for Psychiatry*, 1977).

There are a few excellent reviews and bibliographies that comprehensively treat the physical and mental relationships between pets and man (Corson, 1975; Pet Food Institute, 1976; Friedman, 1978; Arkow 1977). Observers of pet-human interactions and relationships report that the physical, emotional, and social closeness of a pet pose minimum psychologic risk for people. Pets provide relationships in which a person can choose the depth of involvement and can dictate most of the rules with little or no risk of rejection or retaliation (Woolfender, 1977). A dog will love and provide companionship whether a person is tall, short, slender or fat, pretty or ugly, pleasant or difficult. Attachment and loyalty of a pet is unreservedly given, unconditional with no demands, predictable and nonjudgmental, in human terms. Pets provide an unbounded friendship that is a source of warmth and companionship and is less competitive and more dependant than generally experienced in human friendship. Pets exhibit overt innocence, love, trust, and total dependancy. They act as perpetual infants.

The reported evidence accumulating through observation and experience asserts that pets serve as mediators, facilitators, aids, catalysts, and cotherapists in the treatment of several forms of physical as well as emotional illnesses and disabilities. They may provide aid to a level of control and even affect the clinical expression of some of these problems, *but it is recognised that the body of evidence from programs of PFT does not support a view that pets alone can be a therapeutic cure.* The pet can allay anxiety in a clinical setting where patients focus on pets with resultant soothing effect on behavior. Human therapists are aided through pets facilitating the start of therapeutic support by creating a communication bridge of common interest, a span created first between the patient and the pet, then a span joining the patient to the human therapist, upon which to build more pointed discussions that the patient might otherwise consider threatening.

Patients gain opportunities to experience and display responsibility for pets' food, comfort, and complete care. Pets are socializing facilitators and can serve a rehabilitative role as well as act as a medium for emotional stabilization, particularly in persons mildly disturbed (at high risk). Psychotic children and adults have shown marked improvement when introduced to pets after conventional therapies (occupational, physical, recreational), drugs, and shock therapy failed. Some bedridden residents of nursing homes responded by getting out of bed, walking the dogs, and interacting more with other residents and staff.

Their dependent needs, their serving as attentive and responsive companions, and their role as catalysts in person-to-person interactions enable pets to serve as a constant life reinforcement and anchor for people whose life role, life-style, and relationships with people are undergoing significant change. Pets can provide a means of allaying negative psychological manifestations in the aging, helping to prevent social withdrawal and alienation among them. As a focal point for conversation with others, pets are "social lubricators" (Mugford, 1974).

With all of this suggestive evidence in support of pet facilitated therapy as a valid mode in the health care of physically, emotionally stressed, and disabled people, one may ask, "Are therapists prescribing the use of pets?" Health care administrators, professional therapists, and patient care personnel are the team responsible for initiating and operating programs of human health care. Collectively and individually they establish and affect the physical and psychosocial environment of their institution's residents and patients. Are they aware of the health benefits to be derived through enhancement of this environment with the introduction of pet animals?

In assessing the extent to which psychotherapists use pets in their practice, a poll was taken of a randomly selected sample of members of the clinical division of the New York State Psychological Association (Levinson, 1972). Out of 435 persons, 319 responded; 51 percent had recommended pets for home use, 39 percent reported familiarity with the use of pets in therapy, and 16 percent had used pets in their practice. A similar survey was made of the American Psychological Association Division throughout the U.S. (Rice, 1973). Of 190, 29 (15%) reported some use of animals or animal content in therapy.

SOME CONDITIONS AND CIRCUMSTANCES FOR SUCCESSFUL USE OF PFT

1. Rules of operation must be established with close monitoring of health, humane conditions, and tender loving care.

2. Undesirable side effects must be recognized and attended to.

3. Select the type of pet (dog, cat, others) that works best in the specific situation: exceptional pets for exceptional people.

4. Pets can either go with patient when discharged or remain in the treatment of other patients (patient is advised).

5. The therapist must use the pet as a catalyst.

6. The pet can help the therapist through increasing motivation, particularly in long-term cases, helping to overcome "therapist burnout."

7. Health care facilities must assure that pets do not cause problems with the staff, the patients, or with sanitation.

8. Emphasis must be placed on integrating the social and environmental matrix into the treatment strategy of the individual patient.

9. Activity, docility, food and habitat requirements, and potential problems in health and disease are important in selection of the pet.

10. No immunosuppressed patients should be allowed proximity to a pet.

11. Good veterinary medical care of pets should be maintained. Veterinary technicians can teach animal maintenance to residents and nursing staff.

12. The most frequently mentioned matters for attention, mentioned by nurses serving in a geriatric ward with two years of experience with cat mascots, were —

    (a) Maintain hygienic conditions by letting cats out frequently and during meal times to relieve themselves.

    (b) Routine patient handwashing before meals should be observed.

    (c) Cats should be kept away from bedfast patients.

    (d) Cats should be restricted to dayroom areas (Brickel, 1979).

13. With proper controls (quarantine/medical screening; regular thorough veterinary medical care; proper housing, feeding; sanitation for pets and training of pets, staff, and patients) the potential problems of safety and disease posed by the presence of a pet in a program of PFT appear to be minimal or nonexistant. Such problems are unlikely to occur in establishments that adhere to conventional standards of hygiene.

SOME CONDITIONS AND CIRCUMSTANCES FOR FAILURE OF OR OPPOSITION TO THE USE OF PFT*

1. Injury to pet by patient or the reverse.

2. Patient becoming dependent on pet (no greater than risk of dependency on human therapists).

3. Nuisance problems: barking, odors, wastes, care and attention.

4. Public or individual bias against the pets in institutions of human health care.

5. Purported environmental hazards introduced by pet into the institution.

6. The limited perception of or experience with a biopsychological perspective in medical education and practice.

---

*As perceived by administrators and staff of health care facilities.

7. Potential areas of concern (cats):

   (a) Possibility of fleas.

   (b) Cat illnesses.

   (c) Patients who might harm cats inadvertently.

   (d) Keeping patients' handling cats from putting fingers in their mouths.

8. Personnel not trained to work with animals in relationship to patient.

9. Legal constraints in standard leasing agreements that prohibit the keeping of pets in senior citizen building units and in nursing homes.

10. Animal phobias in patients or staff.

11. It is frequently implied and assumed that—

    (a) Allergy to pets is common.

    (b) Dog biting is common.

    (c) Pets are likely to transmit diseases to people.

    (d) Pets are unsanitary and odorous.

12. Patients might interact with pets to the exclusion of people (This was not encountered in one research program examining this matter.) (Corson, 1975).

13. Maintenance and pet care factors are time and money consuming (Where these were estimated retrospectively, these factors proved minimal.) (Levinson, 1970).

THE ANIMAL IN PFT

What about the pet animal and the animal setting in programs of PFT? What follows is an accumulation of experiences drawn from a variety of published statements regarding programs of PFT.

Dogs have been maintained on a hospital psychiatric ward, in an animal ward with pens (not cages). Dogs have not been debarked because this may remove an important element in the pet-patient interaction. Barking in kennels can be controlled by encouraging sleep at night and by feeding and exercising the pet regularly. There are some dogs that do not bark. A trained pet can be taught to respond to verbal and nonverbal signals.

Each breed of dog, bred for desirable traits of behavior, has a strong tendency to express certain behavior in certain situations. Strains within breeds can vary widely in their behavior from the characteristic breed traits. General personality profiles can be determined by selective testing (most successful in five-week-old puppies). Many of the personality tendencies of puppyhood can be modified by socialization, training, and management. Care should be taken when introducing immature animals into a PFT program to assure socialization training (between three and twelve weeks of age) and experiences that avoid the pet's maturing with qualities of perpetual infantilism. Pets can advantageously be selected and bred for affection (particularly cats) for use in programs of PFT.

Dogs for a hearing-ear program are donated or are handpicked at a shelter for intelligence, health, trainability, disposition, and emotional stability. They are between six months and four years old, and small (more suitable for apartment living) (Woolfender, 1977). These dogs are to excel in discriminating sounds and acting upon them appropriately rather than to excel at verbal commands. The objective is not merely a trained dog but a dog that serves a specific purpose.

An experienced dog trainer and teacher of over 1,000 student trainees for use in PFT says, "Our purpose is to provide dogs extensively trained in all varieties of desired behavior and according to the individual situation. Hand and verbal signal-trained dogs will be used in institutions for shut-ins and needy individuals. Many others may benefit by raising their own puppies, properly trained and supervised by themselves" (Woolfender, 1977).

"For physically and mentally handicapped individuals, a dog best suited is said to be medium-sized, short to medium length coat, must learn tolerance of wheelchairs and various pieces of therapy equipment, must ignore noise and be prepared for patient hesitating and waiting at doors. It must retrieve articles of clothing and assist a handicapped reaching and bending to pick up dropped articles; respond to verbal, hand and facial signals; be totally off leash controlled; and stand firmly, as a support, when the patient requires such for dressing, getting from wheelchair to bed, etc."

"For the deaf or autistic, the dog can be of any size and its coat of any length. It must learn hand signal response and constant eye communication. It will bark and nudge the hand in response to phone and doorbell rings or other matters where the attention of the owner is needed. It will respond to the owner's name as well as its own name when spoken by other individuals. It is smoke and fire-trained to wake the owner or get attention in any emergency, responding to sirens, horns, etc."

"For delinquent children's homes, boy's ranches, young people's aid programs of any kind, the preference is for medium to large dogs, with medium to long coats, but above all dogs showing tolerance to unusual situations, non-aggressive at all times, with elasticity and quick recovery in matters of brief mishandling. This type of dog must show total loyalty, be fun-loving, active, with long staying power, able to retrieve and play games and also able to share and give to several people. It must be bright, quick in response."

"For the elderly or for those in nursing and retirement homes, small dogs only, short or long-haired, are recommended. They must be quiet, non-barkers, dependent and constant with a preference for the indoors and indifference to other dogs when outdoors, and a sedate and non-demanding temperament. Tolerance for canes and other walking equipment is essential when the dog is walking at heel."

"For mental health centers and for alcoholics or drug addicts, either individually or in institutions, the recommended dog is medium to large with medium to long hair coat. Physically this dog should require much brushing, and doting care to keep the human subject in close contact. It must be the dependent and 'leaning-on' or 'one-man' type of dog, intelligent and sensitive."

"Puppies and young dogs needing polishing and finish in their training are desirable as part of the therapy project to be undertaken by the patient, with active supervision in class or private lessons, in regular obedience school situations."

"The original training in all the foregoing cases can entail from 4–12 weeks of intensive work prior to placement. This includes background work followed by 'proof' training at the actual location and in the situations the dog must react to in his eventual job." (Woolfender, 1977).

WHAT KINDS OF PFT PROGRAMS ARE THERE TODAY?

The following, a few of the PFT programs in the U.S.A., are impressive yet not exhaustive. They illustrate the diversity of the sponsoring groups as well as of the type and purpose of programs:

1. The Humane Society of Colorado, in Colorado Springs, started in 1973 (reported in 1979) to arrange visits by shelter animals to patients of nursing homes, hospitals, state schools for the deaf and blind, the rehabilitation center, and Medical Retardation Commission. The program involves teenage volunteers.

2. At The American Humane Society (Englewood, Colorado), dogs are trained to detect sounds and to signal their owners (traffic noise, doorbells, smoke signals, babycrying).

3. The College of Veterinary Medicine in Pullman, Washington, through the initiative and creative leadership of its Dean, Leo Bustad, has formed a People–Pet Partnership Council (Bustad, 1979). Among its objectives are—

    (a) The use of animals in therapy for the mentally disturbed.

    (b) Promotion of pets and companionship programs for the elderly, the lonely, as well as for students in dormitories.

    (c) A referral system for area residents to obtain specially trained pets (handy pets for handicapped, hearing dogs).

    (d) Development of a clearing house to provide information on pet programs and to link resource people with persons seeking assistance.

(e) Establishment of a riding program for disabled.

(f) Placement of pets in dormitories to alleviate student loneliness.

4. The Veterinary School of the University of Pennsylvania has been awarded a three-year, $300,000 training grant under the direction of Dr. Aaron Katcher by the National Institute of Mental Health, Division of Manpower Research and Demonstration. The grant supports a program to educate both veterinarians and mental health personnel to the importance of animal companions in human life, the use of animal companions for mentally and physically compromised individuals, and the behavioral/ emotional responses of all animal owners when their animal companion is threatened by injury or disease. Seminars and clinical teaching are conducted for students in veterinary medicine, social work, and medicine, as well as veterinarians, social workers, and psychiatric residents at a Center on the Interactions of Animals and Society.

5. At the Ashton School for Handicapped Children, Salinas, California, training courses are given in canine therapy, physical therapy, and occupational and speech therapy (Woolfender, 1977). The Canine Therapy Program created by Avis Friberg has trained over 1,000 students thus far.

6. San Francisco Society for the Prevention of Cruelty to Animals operates a Pet-A-Care Program that provides veterinary medical care for minimum cost at the Society's Animal Hospital to pets (two per person) owned by persons over age sixty-five (income not over $5,000). The Society is recruiting volunteers to visit the elderly at home and to care for pets when owners are temporarily unable to do so.

7. The Massachusetts Society for the Prevention of Cruelty to Animals, The American Humane Society, and the Junior League of Boston jointly operate a program to place carefully selected dogs and cats in nursing homes

entirely funded by the sponsors. Before placement, animals are neutered and when placed are given one year of routine medical care. Junior League volunteers are trained for pet care, pet therapy, and problems of the elderly. The volunteers introduce the pet therapy concept to administrators and staff at homes eligible to receive pets. Once an animal is placed, the volunteer continues to supervise the project for one year.

8. A course for trainers of hearing-ear dogs is given at the Newberry Junior College Division in Holliston, Massachusetts.

9. The Humane Society of the U.S., in New York City, has launched a new program of "Children and Animals Together for Seniors" (C.A.T's). Senior citizens who adopt an animal from a shelter receive a stipend and/or certificate to defray expenses for pet food and veterinary medical care. Volunteer boy or girl scouts, 4-H, and other junior and senior high school clubs will assist seniors in pet care by walking, grooming, and training to earn school credits, merit badges, or small stipends.

10. Canine Companions for Independence, Santa Rosa, California, trains dogs individually to meet the requirements of disabled persons:

    (a) The service dog—to carry a pack, pull a wheelchair, recover items, perform rescue work, operate switches (lights), and do protection work.

    (b) The signal dog—to alert deaf to selected sounds such as alarm devices (clocks, fire), doorbells, telephone, baby cries, or distress calls.

    (c) The social dog—in hospitals and schools for multi-handicapped, for characteristics of loving warmth and trusting devotion; also to play and wake-up.

11. Therapeutic horseback riding centers are operating where individuals with a variety of physical and mental disabilities experience therapeutic horsemanship (Riding Academy,

Boxford, Massachusetts; Cheff Center for the Handicapped, Augusta, Georgia).

12. Pet therapy at the Lima State Hospital for the Criminally Insane, Lima, Ohio, is a four-year-old program of PFT participated in by 90 out of 400 patients caring for 20 aquaria and 193 pets, including macaws, parrots, cockatiels, canaries, parakeets, doves, rabbits, gerbils, and hamsters. These animals are maintained and cared for by the patients and are kept full-time in the hospital as cotherapists in an unique and effective program coordinated by David Lee, psychiatric social worker. Rules of the hospital were changed to permit resident pets (Lee, 1979).

WHERE ARE WE IN PFT RESEARCH?

In the past year, the following PFT research proposals from United States institutions came to my attention:

1. "The Therapeutic Impact of Pet Animals (dogs, cats) with the Hospitalised Elderly. C. M. Brickel MA, Veterans Administration Medical Center, Sepulveda, CA. Brickel (1978) hypothesizes that the presence of a ward mascot (dog or cat) will be beneficial both to the elderly hospital patients and to their attending staff. He proposes to test this in the presence or absence of pet mascots through the measurement and analysis of patient behaviors, such as; clinical levels of depression latency and frequency of vocalization, and through observation of study participants, and, the measurement and analysis of patient and staff perception of their social environment."

2. Dr. S. S. Robb, Veterans Administration Medical Center, Pittsburgh, Pennsylvania, (Robb, 1979) proposes "The Study of PFT Intervention for the Impaired Elderly." Her study proposes to identify and measure the scope and extent of impact of pet dogs as an intervention in therapy to improve the mental, social, and physical states of elderly people in long-term care facilities; the effect on the nurses' role, on the dogs, and on the cost of maintaining

this proposed therapeutic program.

3. "Study of the Use of Companion-aids [capuchin, organ-grinder monkeys] for the Physically Handicapped," Tufts University, Boston, Massachusetts (Willard, 1979). This is a proposed extension of an ongoing study of an innovative use of two well trained capuchin monkeys. The monkeys perform daily tasks of feeding, fetching, and aiding single owner quadraplegics. The plan is to expand the number of trained skills, to add animals to the program, and to further evaluate owner acceptance and costs.

4. J. Jasper Jacobson, of Indiana University, Bloomington, Indiana, "The Educational Use of Pets with the Handicapped Youth." The plan is to assess the readiness of professionals (educational administrators and teachers of handicapped youth) to use pets in their educational program (Jacobson, 1980).

Published PFT research has not progressed very far from the early studies of Levinson (1976), Corson (1976), Brickel (1979), and Mugford and M'Comisky (1975). More recently there are the published research studies of Brickel and of Friedmann and Associates. Brickel published "The Therapeutic Role of Cat Mascots with a Hospital-based Geriatric Population — A Staff Survey" in 1979. His survey of the staff following two year's experience identified that the cat mascots were observed to stimulate patient responsiveness, gave patients pleasure, enhanced the treatment milieu, and acted as a form of reality therapy — at an insignificant cost and with no important problems identified as a result of the mascot's presence. Friedmann and Associates (1978) reported that pet ownership in itself is related (in some yet unidentified way) to one year survival of patients who had been hospitalized a year earlier for either angina pectoris or myocardial infarction. "A multiple regression analysis was made with one year survival as the dependant and as independant variables: pet ownership and an index of physiological state (diagnosis and number of previous myocardial infarctions plus congestive heart failure + premature ventricular contractions)." The analyses suggest that the relationship between pet ownership and survival is probably independent of differences in physiological state between pet owners and nonowners. "Pet ownership should be further investigated as a therapeutic tool for discharged patients," with these diagnosed conditions.

DISCUSSION:

PFT has tremendous emotional appeal. However, when administrative sanction and permission are sought and funds are needed to support research, the greatest of emotional appeal is not the determining factor in assuring successful acquisition of permission, sanction, or funds.

In so far as gaining administrative sanction and permission is concerned, a recent breakthrough is the passage of a law on April 23, 1979 in the state of Minnesota reading, "Facilities for the institutional care of human beings licensed under Minnesota Statutes 1978 Sec. 144.50 may keep pet animals on the premises subject to reasonable rules as to the care, type and maintenance of the pet" (Seufert, 1980). The decision to have or not to have pets is up to the facility. If the facility opts to have a pet, the Health Departments rules will place restrictions on the maintenance, such as to protect the health, safety, comfort, treatment, and well-being of human residents and to assure that pets will not pose a health hazard or create a nuisance.

SUCH REASONABLE RULES

1. Restrictions will be placed on where pets can be allowed. Example: They will be excluded from food services or preparation areas or in nursing stations or areas where clean linen and sterile supplies are stored.

2. Rules will establish the health facility's responsibility for—

   (a) Maintaining sanitary conditions where pets are located.

   (b) Maintaining good animal health and immunization.

   (c) Ensuring that pets do not create a nuisance to facility residents.

3. The keeping of specific species of pets (turtles, reptiles, psittacine birds) not found freed of potentially communicable human-animal disease(s) will probably be restricted.

When issued, these rules will very likely become a prototype and will bear importantly on the development of similar rulings elsewhere. Fortunately, the Minnesota Department of Health is actively calling upon the knowledge of the Veterinary Medical Health Professionals in developing these rules.

If ever there was a need and time for a handbook of explicit guidelines of ways to establish and operate programs of PFT, it is now. I am aware of a few publications that discuss how to start a pet therapy program. A flyer and an expanded fifty-page report ($1.50) are available from the Humane Society, P.O. Box 187, Colorado Springs, CO 80902 (Arkow, 1977; 1977a).

E. Cooper addresses some aspects of preparation for the use of "Pets in Hospitals" (1976). Dr. Boris Levinson and Rema Freiberger are completing a book *Pets Are Good* that will be published by late 1980 or early 1981 (in press). Written in popular style, it will deal with how to use pets during different stages of human life, including what kinds of pets are best and how to care for such animals. This book will contain an extensive bibliography.

There are PFT related programs that are vigorously directed toward training and educational efforts: Dean Leo Bustad at Pullman, Washington, Dr. Alan Beck at Philadelphia, Pennsylvania, Dr. S. Corson at Ohio State University, Columbus, Ohio, and Dr. M. McCulloch, Delta Foundation, Portland, Oregon are developing urgently needed programs dedicated to the training and education of animal breeders, animal trainers, human-animal technician therapists, veterinarians, physicians, nurses, social workers, and behavioral scientists in PFT. Such training would include psychophysiology, psychosocial environments, behavioral sciences, pet-human interactions, and interrelationships and logistics in establishing and carrying out a prescribed program of PFT. Much is already known about how to train animals for several PFT roles. The program for training dogs as guides for the sightless has developed to the point where the presence of such trained animals is accepted in most all health care institutions. Surely PFT planners could gain know-how by studying and adopting essential practices and highly successful experiences of the dog guide programs. PFT dog trainers in particular have developed new insights.

If one asks whether introducing a pet into a health care facility poses a potential threat to the health of a patient or patients, residents or staff, the answer is a reassuring "No," provided that "no pet animal is introduced that is identified through rigid and thorough screening to be carrying any agent known to cause disease in man." An exception might be the occasional allergic reaction, which is to be avoided. It is known how to select, provide for, and maintain healthy pet animals in human surroundings. Also, many health care institutions with programs of animal model study safely maintain many species of study animals within their main physical facility.

Rigorous prescreening and elimination of known agents of disease of the pet species potentially affecting man as well as strict adherence to practices of sanitation and maintenance that prevent entry of these agents into an institution are the key essentials to healthy pet animals, prior to selection and training for PFT use in a health care facility. This is not to belittle expressed concern for potential human disease with the introduction of pet animals in programs of PFT. Veterinary Medical Health Professionals are prepared to deal with these matters if responsibly involved in a PFT program from the initial planning stages onwards.

There are reliable methods and tests for measuring human satisfaction and mental health improvement. The empirical use of pet animals in health care as facilitators of therapy needs supportive validation by studying and reporting measured results of PFT in clinical situations. It is essential that such clinical reports appear in appropriate, peer refereed professional journals to give medical credibility to the clinical practice of PFT.

REFERENCES

Arkow, Phil: *How to Start a "Pet Therapy" Program.* Brochure from the Humane Society of the Pike's Peak Region, Colorado Springs, CO, 1977.
_____: *Pet Therapy: A Study of the Companion Animals in Selected Therapies.* Booklet from the Humane Society of the Pike's Peak Region, Colorado Springs, CO, June 1977.

Brickel, C. M.: The therapeutic roles of cat mascots with a hospital-based geriatric population: a staff survey." *Gerontologist* 19, no. 4 (August 1979): 368–71.

Brickel, C. M. and Brickel, G. K.: A review of the roles of pet animals in psychotherapy and with the elderly. *International Journal of Aging and Human Development.* In press.

Buckner, Bob, Missouri Veterans' Home, St. James Missouri, to Dr. Michael W. Fox, Institute for the Study of Animal Problems, Washington, D.C. Correspondence, June 7, 1978.

Bustad, L. K.: How animals make people human and humane. *Modern Veterinary Practice* 60, no. 9 (September 1979): 707–10.

Cooper, J. E.: Pets in hospitals. *British Medical Journal* 1, No. 6011 (March 20, 1976): 698–700.

Corson, S. A. et al.: Pet-facilitated psychotherapy. In Anderson, R. S. (Ed.): *Pet Animals and Society.* London, Bailliere Tindall, 1975, pp. 19–37.

Corson, S. A. et al.: Pet-facilitated psychotherapy in a hospital setting. *Current Psychiatric Therapies 15*: 277–86, 1975.

Corson, S. A. et al.: *The Socializing Role of Pet Animals in Nursing Homes: An Experiment in Nonverbal Communication Therapy.* Preliminary draft of a paper for presentation at an International Symposium on Society, Stress and Disease: Aging and Old Age. Stockholm, Sweden, June 14–19, 1976.

Enquist, C. L. et al.: *Quality of Life Report.* Professional Services Letter, IL-11-78-40. Washington, D.C., U.S. Veterans Administration, Department of Medicine and Surgery, July 28, 1978.

Friedman, E. et al.: Pet ownership and coronary heart disease. *Circulation* 58 no. 4, Supp. II (October 1978): 168.

Friedman, E. et al.: *Pet Ownership and Survival after Coronary Heart Disease.* Paper presented at the 2nd Symposium on Pets and Society, Vancouver, B.C., Canada, May 30–June 1, 1979.

Inpatient Program Guide for Psychiatry. Program Guide. Mental Health and Behavioral Sciences Service. G-13, M-2, Part X, 8(c) 26. Washington, D.C., U.S. Veterans Administration, Department of Medicine and Surgery, December 15, 1977.

Jacobsen, Jamia Jasper, Indianapolis, Indiana, to Michael W. Fox, Washington D.C. Correspondence, January 6, 1980.

Lee, David: Birds as therapy at Lima State Hospital for the criminally insane. *Bird World* 1 no. 6, (February–March 1979): 14–16.

Levinson, B. M.: Nursing home pets: a psychological adventure for the patient. *National Humane Review* (July–August 1970): 14–16.

————: *Pets and Human Development.* Springfield, Thomas, 1972.

Levinson, B. and Freiberger, R.: *Pets Are Good.* New York, Richard Marek, forthcoming.

Marcus, C. L.: Animal therapy: a new way to research handicapped children. *NAAHE* (National Association for the Advancement of Humane Education) *Journal* 3 (Summer 1976): 5–10.

Minnesota Legislature, Senate. *An Act relating to health; permitting placement of pets in certain institutions...* Laws of Minnesota, Chapter 38, S. F. No. 307, 71st Legislature, 1979 sess., Minnesota Session Law Service, Laws 1979, p. 53.

Mugford, R. A. and M'Comisky, J. G.: Some recent work on the psychotherapeutic value of cage birds with old people. In Anderson, R. S. (Ed.): *Pet Animals and Society.* London, Bailliere Tindall, 1975, pp. 54–65.

Pet Food Institute: *Literature Review of the Psychological Aspects of Pet Ownership.* Washington, D.C., 1976.

Rice, S. S. et al.: Animals and psychotherapy: a survey. *Journal of Community Psychology* 1 no. 3, (July 1973): 323–26.

Robb, Susanne S., Associate Chief, Nursing Service for Research, Veterans Administration Medical Center, Pittsburgh (UD) PA, to Jules Cass, D.V.M., Chief Veterinary Medical Officer (151E), Medical Research Service. VA Central Office, Washington, D.C. Correspondence, February 21, 1979.

School of Dental Medicine, University of Pennsylvania: *Veterinary Medicine and Human Behaviour.* Description of work supporting training grant "Veterinary Medicine and the Study of Human Behaviour," 1978. Philadelphia, PA.

Seufert, Clarice, Chief, Survey and Compliance Section, Health Systems Division, Minnesota Department of Health, to Dr. Jules Cass, Veterans Administration, Washington, D.C. Correspondence, January 9, 1980.

Wallin, P.: Pets and mental health: Some psychological perspectives on the pet/person relationship and implications for veterinary practice. In *The Newer Knowledge About Dogs: 28th Gaines Veterinary Symposium.* New York, Gaines Dog Research Center, September 23, 1978, pp. 8–12.

Willard, M. J. Assistant Clinical Professor of Rehabilitation Medicine, Rehabilitation Department, Tufts Medical School, Boston, MA 02111. *Study of the Use of the Capuchin Monkey as Companion-aides for Quadriplegics.* (October 1979). Personal Communication.

Woolfender, John: Canine therapists aid the handicapped: dogs open doors to communication. *Monterey (CA) Peninsula Herald Weekend Magazine* (March 13, 1977): 3–7.

CHAPTER 8

COMPANION ANIMALS AS BONDING CATALYSTS IN GERIATRIC INSTITUTIONS

SAMUEL A. CORSON AND ELIZABETH O'LEARY CORSON*

> *The fact that for the last fifteen or twenty years of his life a man should be no more than a reject, a piece of scrap, reveals the failure of our civilization.*
> Simone de Beauvior
> The Coming of Age

THE SOCIOECONOMIC, CULTURAL, AND PSYCHOSOCIAL ISOLATION OF THE AGED IN INDUSTRIALIZED SOCIETIES

The surveys by Simone de Beauvoir (1970) and Butler (1975) documented the extensive degradation and socioeconomic isolation of the aged in industrialized societies. Society's underlying attitudes to old age were described by Butler (1975) as "ageism . . . a process of systematic stereotyping of and discrimination against people because they are old." Obligatory age-related retirement plays havoc with the aged even years before the actual date of forced retirement because many middle-aged individuals may experience a feeling of worthlessness and impending psychosocial disintegration and socioeconomic deprivation in anticipation of obligatory retirement.

The process of systematic exclusion of the aged from the mainstream of economic, social, and cultural life is reminiscent of the psychology of planned obsolescence based on the belief that when products get old they are to be discarded and replaced with new products designed in turn for early obsolescence.

*Supported in part by the Ohio State University Graduate School Biomedical Research Support Grant and USPHS-NIH grant HL 20861.

146

In industrial societies with high population mobility, the emotional trauma of economic and social isolation in the aged is often superimposed on an earlier layer of psychologic stress associated with the "empty nest syndrome." In the case of many middle-aged couples or single parents, children not only leave the home but may move away to distant parts, thus breaking or weakening the chain of mutual social and emotional support. The loss of economic and social roles induced by obligatory age-related retirement thus may accentuate the psychologic stress of the empty nest syndrome and lead to loss of self-esteem and of the ability or willingness to maintain some semblance of independent functioning, socialization, and goal directed activities.

THE DEBILITATING PSYCHOSOCIAL MILIEU OF GERIATRIC CUSTODIAL INSTITUTIONS IN THE UNITED STATES

Deplorable as is the economic and psychosocial status of the elderly in America, the psychosocial milieu of many custodial geriatric institutions and nursing homes is such as to intensify further personality disintegration, alienation, and an infantile type of dependence. The social structure of any custodial institution tends to perpetuate and exacerbate the very deficiencies that brought the residents there in the first place. Thus, a vicious cycle of debilitation, social degradation and dehumanization is established.

The conditions in some nursing homes have been so deplorable as to be labeled by a newspaper as 'warehouses for the dying' (Bevan, 1972). Vladeck (1980) appropriately entitled his book on nursing homes: *Unloving Care: The Nursing Home Tragedy.*

Although only 5 percent of the elderly in the United States live in nursing homes, in actual numbers the nursing home population is quite substantial. According to Kane and Kane (1978), in the United States in 1973 there were 1,327,700 beds in 21,834 nursing homes, as contrasted with 1,030,432 beds in a total of 6,458 general medical and surgical hospitals. The low quality of care in nursing homes can be surmised from the fact that the average number of personnel in nursing homes is 64 per 100 residents, as contrasted with 243 per 100 in hospitals. Moreover, the Kanes point out that nursing home "personnel tend to be less well paid, poorer trained, and less satisfied with their work than those in other parts of the health care system." Another factor that may contribute to inadequate care in nursing homes is the fact that 77 percent of these homes are privately operated as contrasted with hospitals, of which only 13 percent are proprietary. According to Kane, "mental health services are scarce in these facilities" and "heavy doses of psychoactive medication are commonly used."

In summary, the psychosocial structure of a typical nursing home in the United States has the following characteristics:

1. It is essentially a closed social group.

2. It has a low staff/resident ratio, thus making it difficult to individualize treatment.

3. It is a highly regimented social organization, leaving very little room for the retention of a sense of individual responsibility and a feeling of dignity.

4. It is a mass-oriented social organization, leaving very little room for privacy and initiative.

5. The residents tend to lose an important life-sustaining and life-enriching driving force, a sense of purpose and engagement in satisfying goal-directed activities.

6. It fails to furnish an environment conducive to the maintenance and development of positive affective states, a feeling of being needed and respected and a feeling of being loved and an opportunity to reciprocate such feelings.

7. The residents lack socially sustaining tactile contacts.

8. Many of the residents may suffer from varying degrees of sensory deficits, particularly in vision and hearing. These losses contribute to further tactile and social isolation, thus leading to a vicious cycle of social deprivation and psychological-emotional disintegration and disorientation.

A decrease in hearing acuity may have particularly serious consequences in regard to social interactions. A person with significant hearing loss may often be thought of as being confused and irrational because the individual may have heard correctly only a few of the words when spoken to. Consequently, his responses may be irrelevant. After several such incidents of miscommunication, the person may finally give up altogether any attempts at extensive verbal communication, which would naturally lead to further social isolation and the development of serious psychopathology as pointed out by Shore (1976).

The forcible relocation to a nursing home is, in itself, a stressful situation. According to the data published by Holmes and Masuda (1974) and Rahe (1974), many life changes at any age may be pathogenic. Certainly an involuntary (or even a voluntary) relocation to a nursing home represents a stressful and psychologically and physiologically debilitating event.

Added to the insult of such a drastic life change and parting from family, friends, and neighbors, the aged often encounter a further blow of having to part also with a beloved pet animal since many (if not most) nursing homes and other institutions (even many apartment houses or condominiums) will not permit pets. Thus, the last link with satisfying social interactions is broken, and the stage is set for loneliness, depression, and psychologic and physical deterioration.

The object of this chapter is to analyze the role pet animals may play in improving the quality of life of the aged. We will also describe preliminary results of experiments designed to elucidate the socializing role of pet animals in geriatric custodial institutions and the improvement in social interactions between nursing home residents and the staff. We advisedly referred to this form of therapeutic intervention as *pet facilitated psychotherapy* (PFP) to indicate that this is not suggested as a panacea or an exclusive form of therapy but rather as a form of therapeutic facilitation and social lubrication.

THE RATIONALE OF USING COMPANION ANIMALS
IN PSYCHOTHERAPY

I think I could turn and live with animals, they are so placid
  and self-contain'd,
I stand and look at them long and long.
They do not sweat and whine about their condition,

They do not lie awake in the dark and weep for their sins,
They do not make me sick discussing their duty to God,
Not one is dissatisfied, not one is demented with the mania of
owning things,
Not one kneels to another, nor to his kind that lived thousands
of years ago,
Not one is respectable or unhappy over the whole earth.

So they show their relations to me and I accept them,
They bring me tokens of myself, they evince them plainly in
their possession.

Walt Whitman
"Song of Myself"

One significant component of PFP is the utilization of positive, reassuring nonverbal communication signals between carefully selected and well-trained pet animals (particularly dogs) and socially withdrawn, self-effacing, and depressed humans. As long ago as 1872, Charles Darwin pointed out the importance of nonverbal communication in his book *The Expression of Emotions in Man and Animals* (Darwin, 1965).

Experienced physicians have utilized nonverbal communication signals as important aids in the diagnosis of physical and emotional illness. A major stimulus to the elaboration of the concept of "stress" by Hans Selye (1936, 1971) was his observation of the facial expressions and overall appearance of patients suffering from a variety of diseases in a hospital ward. E. H. Hess (1975) reported on his extensive observations of nonverbal communication as represented by changes in the pupillary diameters.

One important aspect of nonverbal communication has been largely neglected. We are referring to negative nonverbal signals that sick and old people often receive from well people, including physicians and other health professionals. Even with the best training and the noblest motives and the kindest intentions, healthy people tend to send negative nonverbal signals to the sick, the infirm, the aged, and the mentally and physically retarded. This may lead to a vicious cycle of social isolation, mistrust, suspicion, and ego disruption. This, in turn, may stimulate a chain of physiologic and psychosocial events that may exert further debilitating effects on the patients and the patient-staff interactions and hinder the restoration of homeostatic mechanisms essential for recovery and well-being.

Deleterious influences of negative nonverbal communication signals are likely to be particularly distressful in institutional settings, especially in psychiatric hospitals, nursing homes, and other custodial institutions where the ratio of staff to patients is low and the extent and intensity of negative emotions is likely to be high and overwhelming, thus leading to institutionalized loneliness and emotional starvation.

One method of breaking this vicious cycle of mutual negative reinforcement is the introduction of carefully selected, well-trained pet animals. Animal pets, particularly dogs and cats, have long played a humanizing and socializing role in human interactions. Records from Egyptian and Chaldean monuments suggest that several breeds of dogs were developed as early as about 3000 BC. Davis and Valla (1978) reported evidence for a dog-human companion relationship as early as 10,000 BC during the Epipaleolithic period. This evidence is based on three canid finds from the Natufian in the northern Israeli sites of Ein Mallaha (Eynan) and Hayonim Terrace. One of these finds includes a dog puppy skeleton buried with a human. The authors conclude that "the puppy, unique among Natufian burials, offers proof that an affectionate rather than a gastronomic relationship existed between it and the buried person, an addition to our knowledge of the way of life of Natufian hunter-gatherers."

The fact that so many people in Western countries own pet animals would suggest that these animals must serve some important social functions. In the United States the dog and cat population has been estimated at about 100 million animals. The monetary expenditure for the purchase and maintenance of dogs and cats in the United States has been estimated at around five billion dollars per year. Some seventy million dollars were spent in 1974 for television advertising of pet foods (Bureau of the Census, 1975). Szasz (1968) reports cases of extravagant spending on pet animal maintenance.

Thus the statistical data and the economic costs of pet animal ownership alone suggest that these animals must serve some significant social and companionship functions since in most cases these pet animals do not have economic value. The development of the dog-people partnership has been depicted extensively by Lorenz (1964), Scott and Fuller (1965), Levinson (1969), and Fox (1971).

The attachment humans develop for pet dogs may be related to two prominent qualities of many dogs: their ability to offer love and tactile reassurance without criticism and their maintenance of a sort of perpetual infantile innocent dependence, which may stimulate our natural tendency to offer support and protection.

The important role played by tactile contact in emotional and physical well-being and in social interactions in humans and in other higher organisms was elegantly documented by Ashley Montagu (1978) in his classic book *Touching*. Montagu points out that touch is the earliest sensory system to become functional. He therefore argues that the psychobiological significance of touch may be subsumed from the general embryological law that states that the earlier a function develops the more fundamental it is likely to be. Touch sensation, therefore, represents significant afferent somatopsychic input essential for optimal physical, emotional, and mental health. Montagu cites many examples from experimental studies in young animals demonstrating that the "proper kind of cutaneous stimulation is essential for the adequate organic and behavioral development of the organism." He reports that touching as a means of gentling young rats leads to better adjustment of the adult rats to stressful situations. Montagu cites anecdotal reports on dolphins, who like to be gently stroked. Two other anecdotal reports are intriguing, although Montagu does not document them: One anecdote relates that "hand-milked cows give more and richer terminal milk than machine-milked cows." This fits in with the common experience that many domestic animals like to be petted and that such petting is apparently pleasurable both to the animal and to the human. The other report suggests that "child-battering and abusing parents, who were themselves neglected and abused as children, rarely report having had a childhood pet."

Harlow (1974), on the basis of his extensive, well-controlled studies on baby monkeys, pointed out the importance of tactile comfort for the biological and psychosocial development of these primates. In some of these experiments the infant monkeys were provided with surrogate wire or cloth covered "mothers." In one experiment, four newborn monkeys were provided with four surrogate cloth covered "lactating" mothers and four nonlactating wire mothers. For four other infant monkeys the wire mothers were "lactating," while the cloth mothers could not offer milk to the infants. The remarkable findings were that in both experiments *the infant monkeys preferred to cuddle with the softer cloth surrogate mothers, regardless of the opportunity of receiving milk.* Harlow concludes that "it is clearly the incentive of contact comfort that binds the infant affectionately to the mother. If derived drive is to be involved as an explanation, this drive must be fashioned from the whole cloth rather than whole milk."

On the basis of his pioneering studies on conditional cardiac responses in a variety of dogs (using electrocutaneous reinforcement), Gantt (1972) presented an insightful analysis of the Effect of Person on heart rate. Patting on the head in most dogs led to a marked bradycardia and slower and deeper respiration. Gantt and his collaborators also demonstrated that canine tachycardia (in response to a loud bell) could be converted into bradycardia when the bell was used as a conditional stimulus for petting. In one catatonic dog, petting decreased the systolic blood pressure from a value of 150 to 80. Gantt concludes that the Effect of Person "can have tremendous influence in therapy, possibly through the application of its tactile component."

One of Gantt's students, Lynch, and his collaborators observed in a sample of 225 coronary care patients a significant reduction in ventricular arrhythmias following a simple tactile procedure like pulse palpation (Lynch et al., 1977).

In our own studies on certain breeds of dogs that exhibited marked and sustained conditional tachycardia in a Pavlovian conditioning room where aversive reinforcement had previously been applied (Corson and Corson, 1976), we repeatedly observed that petting the dogs led to dramatic amelioration of this psychogenic tachycardia.

Apart from tactile comfort and reassurance, pet animals also offer an opportunity (and a challenge) for adults and the aged to engage in childlike play. Such "regressive" play activities may have significant stress-reducing and rejuvenating influences.

Pet animals have been used in psychotherapy on and off for many years, going back to the eighteenth century (Levinson, 1969). On the basis of a survey among clinical psychologists in New York state, Levinson (1972) reported that 33 percent of the respondents used pets in their therapeutic procedures, with dogs being the preferred animal.

Under the direction of Dr. Stuart M. Finch and head nurse Alice Williams, the University of Michigan Children's Psychiatric Hospital had a dog, Skeezer, live on the ward for some seven years as a sort of cotherapist (see Yates, 1973). However, no systematic research was conducted on this form of therapy.

The most extensive use of pets in the psychotherapy of children was reported by Boris M. Levinson (1969, 1972) in two instructive and elegantly written monographs. The current State of the Art is succinctly summarized by Levinson in his 1972 book: "The survey reflects considerable confusion and lack of direction concerning pet-aided psychotherapy. . . . We need highly trained imaginative and extremely rigorous research to establish principles and boundaries in the use of pets in psychotherapy. The outcome of such therapy must be evaluated in terms of the needs of each individual patient. . . . Much must be learned about how to train dogs for special psychotherapeutic work with both children and adults."

Brickel (in press) conducted a survey of the 19 nursing staff in a total-care hospital geriatric ward where two cat mascots had been maintained for a period of two years. Most of the patients had a diagnosis of chronic brain syndrome. Brickel concluded that the feline mascots "were observed to stimulate patient responsiveness, give patients pleasure, enhance the treatment milieu, and act as a form of reality therapy." Ten of the nurses reported that the presence of the cats was helpful also to the staff and staff morale. Only three nurses were not in favor of cat mascots.

We previously described our experience with the use of companion animals in psychotherapy in a psychiatric hospital (Corson et al., 1975a; 1975b; 1977; Corson and Corson, 1979) where we conducted a systematic study of pet animal-patient-staff interactions, utilizing primarily well-trained dogs and some cats. Of the fifty patients studied, three did not accept a dog. Since we had a limited budget and a small staff, we did not try to discover whether these three patients might have interacted in a positive way with other animal species. Of the forty-seven patients in our investigation, all benefited from interactions with the companion animals. Some did so in a dramatic way. The improvement occurred in regard to individual behavioral patterns as well as in regard to social interactions with other patients and with the staff. There was a general heightening of the *esprit de corps* of the entire unit where the animals were introduced.

Because of the difficulties inherent in the evaluation of the efficacy of any form of psychotherapy, we selected for our in-depth studies chiefly patients who had been in the hospital for a relatively long time and had failed to respond favorably to traditional forms of therapy. We postulated that if the addition of PFP led to improvement in some of these patients, this would suggest that PFP had some therapeutic value.

Our controls, however, were deficient because for economic and ethical reasons we were not able to eliminate other forms of therapy while the patients were exposed to pet facilitated psychotherapy. Actually, this was not a serious defect in our experimental design since *PFP was not proposed as a substitute for other forms of therapy but as an adjunct to facilitate the resocialization process.* When appropriate research funding becomes available, it would be desirable as the next best control to include a larger number of patients at an equally low level of psychological functioning and offer PFP to only 50 percent of such patients in a random selection design.

However, it is not feasible to utilize such an experimental design with PFP. The introduction of a pet animal to one patient in a given ward invariably involves the other patients, not only in this particular ward but in the entire hospital. Word about the companion animal soon spreads throughout the entire institution. Therein, of course, lies the socializing value of PFP.

Therefore, appropriate control studies with PFP would have to involve two or more comparable psychiatric hospitals; pet animals would be introduced to patients in only one hospital, the other hospital being used as a control.

PFP IN A NURSING HOME

If you pick up a starving dog and make
him prosperous, he will not bite you.
This is the principal difference between
a dog and a man.
Pudd'nhead Wilson
Pudd'nhead Wilson's Calendar

In the summer of 1975 we were forced, due to lack of research funds, to disband our dog colony at the Ohio State University Psychiatric Hospital. Through the good offices of Dr. Thomas Graham of Massillon State Hospital an offer came from the late Mr. Donald DeHass, Administrator of Castle Nursing Homes in Millersburg, Ohio, to take most of our dogs to use as pet cotherapists.

Mr. DeHass had long been a dog owner and breeder, and he was convinced of the benefits of pet companionship to persons of all ages; his Doberman pinscher, a well-tempered, well-liked dog named Blue, regularly made rounds with him.

We undertook a research program at Castle Nursing Homes, with the aim, among other things, of quantifying patient responses to PFP and of investigating effects of PFP on staff morale and performance. We gave considerable thought to the fact that the plight of the institutionalized old person is in several ways even more difficult than that of the old person living alone in his or her home.

THERAPEUTIC PROCEDURES IN A NURSING HOME

Fox terriers are born with about four times
as much original sin in them as other dogs.
Jerome Klapka Jerome
Three Men in a Boat

Types of Pet Animals Best Suited for PFP

As mentioned earlier, we primarily utilized dogs. These animals had been conditioned and trained in our laboratory or by Donald DeHass at the Castle Nursing Homes, so we could reasonably predict the behavior of these dogs under different conditions to the extent that behavior in general can be predicted in animals or man.

One of the reasons for the success of our PFP program was the fact that we had a variety of dog breeds with different temperaments and different behavioral repertoires and modes of emotional expression. It was thus possible to try to match the particular needs of a particular patient or nursing home resident with an appropriate dog.

We have used the following breeds of dogs in PFP: wirehair fox terriers, Border collies, poodles, Labrador retrievers, Doberman pinschers, cocker spaniels, dachshunds, and some mongrels (German shepherd X husky hybrid).

With individuals who are physically or psychologically bedridden or are largely limited to a wheelchair, small dogs appear to be especially useful. In particular, well-trained wirehair fox terriers and small poodles (miniatures, toys) were especially appealing to love-hungry, depressed, withdrawn residents or to those with low mobility. These dogs, with their resilience, aggressive friendliness, good humor, and playfulness, served as effective icebreakers and social matchmakers.

Methods of Introducing Dogs to Residents

Castle Nursing Homes is a privately owned institution at Millersburg, Ohio, situated about 130 kilometers from the Ohio State University at Columbus. It has accommodations for about 800 residents, including the aged as well as individuals of different ages with various degrees of mental or physical disabilities. The housing units vary from cottages (suitable for two to six individuals) and apartment buildings to units designed as skilled nursing facilities for residents requiring long– or short-term nursing and medical care.

The introduction of PFP was always preceded by a discussion with the residents regarding their attitudes towards animals and their possible likes or dislikes for different kinds of pets. In the case of patients who were physically or psychologically bedridden or restricted to a wheelchair or disoriented, a small adult dog or a puppy was brought to the resident. The pet-resident interaction was always under the supervision of a staff member who was thoroughly familiar with the behavior of the animal and who had good rapport with the resident and the pet dog.

In the case of residents living in a dormitory-type ward, a large cage with weaned puppies was placed in the foyer. A nurse or pet therapist demonstrated to the residents how to handle and feed the puppies, change the papers, etc. Generally, some residents would voluntarily and spontaneously "adopt" a puppy and take the major responsibility for its feeding and for maintaining the cage in good sanitary condition.

Residents living in small cottages preferred to have a large dog living in a doghouse placed near the cottage. The dog helped to transform the cottage residents into a sort of extended family.

In some cases we found it useful to incorporate PFP as a component of a token system, using a pet dog as a reward in a behavior modification program. We have found this system useful in our studies with hospitalized psychiatric patients, particularly with adolescents.

Methods of Recording the Progress of PFP

A questionnaire was developed and incorporated into nurses' notes. These included observations on the physical and emotional well-being of the resident, social interactions with other residents and staff, changes in personal hygiene and appearance, changes in medication, etc. The questionnaire was constructed on the basis of a 10-point scale so that all the nurses needed to do was circle or check the appropriate items (*see* Appendix 1).

The most useful instrument was videotape recording of the resident-pet-staff interactions. These recordings permitted the quantification of the verbal and temporal parameters of the residents' social interactions and the effects of the pet animals on the tactile and other forms of nonverbal communication patterns. In many cases, the videotape recordings also served as therapeutic feedback for the residents.

Therapeutic Results in a Nursing Home

Pet animals, especially carefully selected, well-trained dogs, offered the residents of a nursing home a form of nonthreatening, reassuring nonverbal communication and tactile comfort and thus helped to break the vicious cycle of loneliness, hopelessness, and social withdrawal. Pet animals acted as effective socializing catalysts with other residents and staff and thus helped to improve the overall morale of the institution and create a community out of individuals, many of whom were separated, detached, unhappy, and self-pitying.

The introduction of PFP served as a form of reality therapy and helped to transform dependent, infantilized, self-neglecting behavior into responsible, more self-reliant modes of interaction. In many cases pet dogs involved the residents in walks and running bouts, thus helping to improve the physical and emotional status of the aged. One overweight resident lost thirty pounds during a period of about a month after she had become involved with PFP. Walking and running with a dog eventually led to a general increase in physical activity and overall well-being. Moreover, walking or running with a dog tends to broaden social interactions and expand the geographic horizons of the residents.

In some cases the pet dogs helped to involve the aged in other rejuvenating activities. For example, one resident, who had been living in the nursing home incommunicative and self-isolated for some twenty years, shortly after being introduced to a pet dog began to draw and paint pictures of dogs and show off his artistic creations to staff, other residents, and visitors. Eventually the walls of his room were covered with drawings of various types of dogs, some of the drawings demonstrating a great deal of skill and perception (*see* Case 1).

PFP WITH MENTALLY RETARDED INDIVIDUALS

The Castle Nursing Homes housed a number of mentally retarded males and females. PFP was introduced to several such individuals. A great deal of careful supervision and training was required at first. Gradually some mentally retarded individuals took over responsibility for the animal care. Thus far we have found that with some mentally retarded females, puppies were the most useful animals. Some of these residents became very much attached to the puppies, expressing the hope that they could keep these animals "as long as I live." They even talked about eventually breeding these dogs so they could have new puppies.

PFP AS A FORM OF EMOTIONAL SUPPORT TO RESIDENTS WITH TERMINAL DISEASES

We had an opportunity to observe at the Castle Nursing Homes the beneficial effects of pet dogs on two mentally retarded sisters, one of whom (Flora, thirty-six years old) was a victim of terminal cancer. She spent most of her time in bed, disinterested and lethargic. The introduction of a small dog not only got Flora out of bed but even encouraged her to take the dog for walks, first in the building and later outdoors.

When her younger sister Dolores (thirty-two years old) learned that Flora was to be taken to the hospital, perhaps never to return, she asked for a dog as a companion substitute for her sister, to whom she was closely attached.

PFP IN PENAL INSTITUTIONS

Shortly after our successful experience with PFP in a psychiatric hospital and a nursing home, we suggested to several criminologists that it might be worthwhile to initiate some pilot studies on the possible socializing and humanizing usefulness of pet animals in penal institutions.

We were, of course, aware of the possibilities of cruelty to animals by some prisoners. Nevertheless, it seemed to us that a well-designed and carefully supervised PFP research project would be well worth the trouble, particularly in view of the fact that thus far few, if any, of the many prisoner rehabilitation programs have proven successful.

Although the criminologists we contacted agreed that it would make sense to initiate some PFP pilot studies at prisons, thus far we are not aware of any such research programs.

In 1975 David Lee, a psychiatric social worker at Lima State Hospital in Ohio, introduced pet animals at that institution with the support of William M. Balson, Superintendent, and Dr. John Vermeulen, Commissioner of Forensic Psychiatry.

Lima State Hospital is a maximum security forensic institution housing about 400 patients, including psychopaths and sex offenders.

The Lima program was not set up as a research project but as a practical aid to improving the social interactions and the overall behavior of the patients. The program was initiated with the introduction of one aquarium and eventually led to the acquisition of some 20 aquaria and 160 pet animals, including parrots, gerbils, rabbits, hamsters, guinea pigs, and a deer. Of the 400 patients, 90 have participated in the program, involving 15 of the 22 hospital wards.

Eventually pets were also introduced into the psychopathic ward. In spite of the fact that some 50 percent of the psychopaths had a history of animal abuse in their childhood or adulthood, David Lee reports that many of the patients are giving humane and loving care to the animals and have demonstrated improved social interactions with their peers, as well as with the staff.

On occasion David Lee makes his ward rounds in the company of a lovely well behaved mixed German shepherd dog (who has a regular staff identification tag on his neck!). We had occasion to accompany Mr. Lee on one of these ward rounds. It was a pleasant and reassuring sight to see and hear the patients from all directions joyfully greet the dog (and us, indirectly, as friends of the dog!). It was hard to believe that this was a maximum security forensic hospital.

The PFP program at Lima would certainly suggest the desirability and feasibility of developing well-controlled PFP studies in other forensic hospitals and in penal institutions, including long-term follow-up programs. The emotional and tactile contact with loving, nonjudgmental pet animals may serve as defusers of alienation and hostility and lead to more adaptive social interactions.

PROBLEMS TO BE CONSIDERED IN INTRODUCING PFP IN CUSTODIAL INSTITUTIONS

Possible Intensification of Psychosocial Pathology by PFP

Before we introduced pets to residents or patients, we were concerned that some individuals might become involved with their pets to the exclusion of interactions with people and that even more pronounced psychosocial pathology might evolve. In the initial stages of PFP, some people did relate exclusively to their pets, but the relationships soon lost their closed-circle aspect. The pets gradually began to serve as catalyzing social links and had a positive effect on other residents or patients. The interest expressed by many patients or residents and the warmth of the staff's response provided the individuals involved with PFP with a widening circle of approval and social interaction.

Possible Cruelty to Pet Animals

An additional concern was that the "security by rank" phenomenon might lead to cruelty to pets—that patients or residents might act out on their pets the indignities they themselves may have suffered at the hands of more powerful persons. Cruelty would not, of course, have been allowed, but it never became a problem in the nursing home or the psychiatric hospital. At this time we have no way of determining whether the lack of cruelty was natural or due to the structure and supervision by the staff.

At the Castle Nursing Homes, some mentally retarded residents had to be carefully supervised in order to avoid a possible injury to the pet dog. This danger was not due to cruelty but to low intelligence and poor coordination. We noticed that the less mentally retarded residents would often help the more retarded to handle the pet carefully and gently.

Problems of Public Health and Sanitation

Dogs and other animals used in pet therapy should first be carefully examined and immunized by a veterinarian. Naturally, continuing systematic veterinary supervision is essential for the success of such a program.

For pet dogs to have the maximum beneficial therapeutic effect on an institution, it would be useful to have permanent, sanitary kennels close to the patient wards or nursing home residence. Dogs should be maintained not in cages but in well-constructed runs or pens with dimensions not less than 8 x 4 feet. Waterproofed walls and floors with drains make it easy to flush the runs daily with hot water and appropriate safe cleaning agents. Each run should be provided with a properly designed wooden box where dogs could sleep. This helps the dogs to keep themselves clean. The dog should be taken outdoors to urinate and defecate before being taken to the patient wards or nursing home residents; if this procedure is followed, "autonomic accidents" can be virtually eliminated.

Adequate bathing and grooming facilities should be provided in the kennel area. Many institutionalized individuals enjoy bathing and grooming the animals. In fact this activity may represent a significant component of the therapeutic usefulness of the pets.

Possible Dog Bites

This problem could be essentially eliminated by two approaches: first, the careful selection, screening, and training of PFP dogs for temperament, reliability, and types of emotional responses; second, careful supervision and education of the residents, patients, and staff in handling the dog and careful observation in the initial stages of PFP of resident/pet compatibility. It is essential for the success of this kind of program to have available a person who is thoroughly knowledgeable about dog behavior and dog training and who has a natural love for dogs and people.

Allergic Reactions to Animals

Possibility of allergic reactions should be carefully checked beforehand. A person would of course not be a candidate for PFP with an animal species to which he is allergic.

CASE HISTORIES OF OLD PEOPLE IN A NURSING HOME
Case 1: Jed

Jed was in his late seventies and had been a nursing home resident for twenty-six years. He was admitted to Castle Nursing Homes, Millersburg, Ohio, in 1949, after suffering brain damage in a fall from a tower. At the time of his admission he was believed to be deaf and mute as a result of his accident.

Figure 8-1.   Jed as he appeared at the first interview.

Through the ensuing years Jed was antisocial and often appeared to be unaware of those who cared for him. The staff used hand signals in attempting to communicate with him. Since Jed could read, the nurses would also write notes to him in order to try to ascertain his needs. Jed's only form of verbal communication was grunting and mumbling, which was incoherent and not necessarily used to make his needs known. He spent most of his time sitting in silence, apparently deaf, with intermittent outbursts of mumbling to himself.

In 1975, shortly after the arrival of the "feeling heart" dogs at the Castle Nursing Homes, the administrator, Donald DeHass, brought dog Whiskey (a German shepherd X husky) to visit Jed. Jed's reaction was immediate—he spoke his first words in twenty-six years: "You brought that dog." Jed was delighted and chuckled as he petted the dog.

With the introduction of the dog, the communication barrier was broken. Jed started talking to the staff about "his" dog. The nurses noted an improvement in Jed's disposition and in his interactions with the staff and other residents. Jed started drawing pictures of dogs and now has a large collection of his canine drawings. This artwork is fairly sophisticated and his collection contains various breeds of dogs. Many of the pictures exhibit well-drawn detail (such as teeth, whiskers, toenails, eyelashes, etc.) and interesting postures of the animals. Jed was later introduced to Fluff, an outgoing, active wirehair fox terrier. Jed immediately made friends with the dog and told her, "I'll take care of you." This

Figure 8-2.   Jed accepting dog with expression of joy and tenderness.

Figure 8-3.   Xerox copy of first painting (original in color) made by Jed of dog Whiskey.

was the beginning of a continuing friendship. The staff and nurses have stated they believe that pet facilitated psychotherapy made Jed a happier person and more willing to interact with people. They also believe that his speech improved, particularly in clarity. Jed made a greater effort to be verbally understood, especially when he wanted to visit his canine friend or to inquire about the dog's welfare.

Case 2: Verna and Madge

Verna and Madge were roommates at Castle Nursing Homes. Verna was in her late seventies and diagnosed as senile. Madge was seventy-nine years old and had been admitted with a diagnosis of senile dementia.

Both patients had to be confined to restraining chairs; they were aggressive and would have violent outbursts, particularly toward each other but also toward the staff. The staff took precautions to keep the ladies separated during the day, a physically safe distance apart. If the ladies were taken out of the restraining chairs while in close proximity, they would immediately start to scream and physically attack one another. One of the nurses understated the situation when she said, "Their personalities clash."

These two patients were introduced to pet facilitated psychotherapy simultaneously in a large and sunny dayroom where other residents were present. Verna and Madge were in their restraining chairs, side by side, when Fluff, a wirehair fox terrier, was brought to them. Verna's high-pitched, complaining singsong ceased immediately. A series of interactions ensued, involving competition, rejection, curiosity, and interest. Sometimes both ladies would pet the dog; at other times, one of the ladies would reject the dog while the roommate petted the animal. Some of the other residents showed a distinct interest in the transactions and enjoyed the session very much. The nursing staff believes that although the pet did not cause any detectable personality changes at the individual level, the dog was the agent in creating improved social interactions *between* the two ladies. After Verna and Madge had a few sessions of PFP together with Fluff, they became more tolerant of each other.

It is interesting to note that the ladies could be taken out of their restraining chairs while Fluff was present and be civil to one another. However, if Fluff was *not* present when the restraining trays were removed, Verna and Madge resumed their hostile attitudes.

Figure 8-4.   Resident, Beverly, in her late seventies appeared depressed and was not responsive to her environment. Her answers were generally unrelated to the questions asked.

Figure 8-5.   The same resident in the same session as in Figure 8-4, interacting tenderly with dog Saxophone, a tea cup poodle.

The staff stated that with continued PFP sessions it might have been possible for Verna and Madge to be tolerant of one another with *and* without Fluff present. However, this form of therapy could not continue, as Verna died. Madge continues to see Fluff on occasion and appears to enjoy playing with the dog. The ward nurse believes that in the case of Verna and Madge "the dog was a tranquilizer."

SUGGESTIONS FOR FUTURE RESEARCH

The results of our preliminary studies on the effects of PFP in custodial institutions, including adult and adolescent psychiatric patients, mentally retarded individuals, and the aged, certainly warrant further systematic long-term investigations. In particular, the following studies are suggested—

1. Institute better controlled studies. As is true for the evaluation of any new therapeutic tool, it is difficult to identify the precise factors that contribute to the improvement of a patient or a resident of a custodial institution. Because of the lack of precision in diagnostic procedures and the great variability in diagnostic categories in custodial institutions, it is difficult to select experimental and control groups that would be comparable. Therefore it appears more appropriate to utilize in such studies an ipsative (intraindividual) method with a longitudinal process oriented design, using each individual as his or her own control. This involves long-term follow-up studies that are possible only with long-term research funding. It would be desirable to arrange for a cooperative consortium-type longitudinal program involving several similar custodial institutions and introducing pet animals in only 50 percent of those institutions.

As mentioned earlier, it is not feasible to select at random residents in a nursing home for PFP or for control studies because the introduction of companion animals to one resident soon tends to involve many other residents and eventually the entire institution. Therein lies the therapeutic and prophylactic value of PFP.

2. In addition to recording effective, behavioral, and social interactive parameters, it would be useful to simultaneously intro-

duce psychophysiologic measurements. This could be done with the aid of microradio-telemetry instrumentation that could be placed on the human subjects as well as on the pet animals.

3.   Determine to what extent PFP could help to decrease excessive reliance on psychotropic drug therapy.

4.   Determine to what extent pet animals could be useful to the aged with sensory deficits.

5.   Investigate the extent to which PFP may help to transfer individuals to more independent forms of living.

6.   Investigate the possibilities of utilizing pet animals as catalysts for introducing into an institution a variety of enriching humanizing activities, such as gardening, music, dance, drama, art, reading, poetry therapy, sports, and other play activities. PFP could include barn animals, such as sheep, goats, horses, etc.

7.   Investigate the extent to which pet animals may help to bridge the generation gap. For example, it would be worthwhile to investigate the feasibility of building well-designed, sanitary animal shelters, e.g. dog pounds, in proximity to nursing and retirement homes, day care centers, elementary and high schools, colleges, and schools for exceptional children and the disabled. This would help to develop mutually beneficial and enriching intergenerational and intercultural communication and cooperative activities.

8.   Investigate the feasibility of utilizing PFP to humanize forensic hospitals and penal institutions.

SUMMARY AND CONCLUSIONS

We previously reported remarkable and sustained positive responses to pet facilitated psychotherapy (PFP) in forty-seven out of fifty hospitalized psychiatric patients who had failed to respond adequately to traditional forms of therapy. This presentation summarizes our observations on PFP (utilizing a variety of carefully selected trained dogs) in an 800 bed nursing home, containing facilities for skilled nursing care for mentally retarded individuals of various ages in apartment buildings and cottages suitable for two to six residents.

The dogs were maintained in large outdoor kennels or in dog

houses kept near the cottages. In the case of dormitory-type wards in skilled nursing facilities, a large cage with weaned puppies was placed in the foyer. The progress of PFP was documented by videotape recordings and a ten-point scale questionnaire scored periodically by the nurses.

PFP is not a panacea and is proposed as an adjunct to, and not as a substitute for, other forms of therapy. Preliminary observations suggest that PFP might help to decrease excessive reliance on pharmacotherapy. Pet animals, and especially dogs, offered nursing home residents (including mentally retarded individuals) a form of nonthreatening, nonjudgmental, reassuring nonverbal communication and tactile comfort and thus helped to break the vicious cycle of loneliness, helplessness, and social withdrawal. Pet animals acted as effective socializing catalysts with other patients, residents, and staff and thus helped to improve the overall morale of the institution and create a community out of detached individuals. PFP served as a form of reality therapy and facilitated the transformation of dependent, infantilized, self-neglecting behavior into more self-reliant modes of interaction. Often the pet dogs stimulated the nursing home residents to engage in a variety of creative pursuits and in walks and running bouts and play, thus leading to a general increase in physical and social activities and general well-being. Further, controlled longitudinal studies are warranted, including the feasibility of introducing PFP in institutions for mentally and physically retarded, the terminally ill, and forensic and penal facilities. Long-term funding is essential in order to permit systematic follow-up observations on possible long-term benefits of PFP.

APPENDIX 1

Nurses' Rating Scales used by Samuel A. Corson, E. O'L. Corson, and R. Gunsett in a Pet Facilitated Psychotherapy Study in a Nursing Home

PATIENT'S NAME: _____

Date: YR _____   MO _____   DA _____   Day of Week _____

PLEASE RATE THE PATIENT ON THE FOLLOWING SCALES:

1. Patient's appetite

    (very poor)   (very good)
        1 2 3 4 5 6 7 8 9 10

2. Patient's general activity level

    (very little)   (very active)
        1 2 3 4 5 6 7 8 9 10

3. Patient's interaction with staff

    (very little)   (very much)
        1 2 3 4 5 6 7 8 9 10

4. Patient's interaction with other residents

    (very little)   (very much)
        1 2 3 4 5 6 7 8 9 10

5. Patient's violent reactions

    (few)   (many)
        1 2 3 4 5 6 7 8 9 10

6. Patient's exercise (regular walks, jogging, etc.)

    (none)   (often)
        1 2 3 4 5 6 7 8 9 10

7. Patient's verbal communication with staff and residents

   (none)   (talks all the time)
   <u>1 2 3 4 5 6 7 8 9 10</u>

8. Patient asks for pet

   (never)   (frequently)
   <u>1 2 3 4 5 6 7 8 9 10</u>

9. Patient's time spent grooming self

   (none)   (grooms self all the time)
   <u>1 2 3 4 5 6 7 8 9 10</u>

10. Patient's dependency on staff (dressing, toilet, etc.)

    (very little)   (totally dependent)
    <u>1 2 3 4 5 6 7 8 9 10</u>

11. Patient's willingness to participate in organized activities

    (never willing)   (very willing)
    <u>1 2 3 4 5 6 7 8 9 10</u>

12. Patient's general level of cooperation

    (none)   (very co-operative)
    <u>1 2 3 4 5 6 7 8 9 10</u>

General observations or comments (please note any personality changes or change of lifestyle):

REFERENCES

Beauvoir, Simone de: *The Coming of Age.* New York, Warner, 1970.

Bevan, W.: On growing old in America. *Science 177* (4052:839, 1972.

Brickel, C. M.: The therapeutic roles of cat mascots with a hospital-based geriatric population: a staff survey. *The Gerontologist*, in press.

Bureau of the Census: *Statistical Abstract of the United States.* U.S. Department of Commerce, Washington, D.C., 1975.

Butler, R. N.: *Why Survive? Being Old in America*. New York, Harper & Row, 1975.

Corson, S. A. and E. O'L. Corson: Constitutional differences in physiologic adaptation to stress and distress. In Serban, G. (Ed.): *Psychopathology of Human Adaptation*. New York, Plenum Press, 1976, pp. 77–94.

Corson, S. A. and E. O'L. Corson: Pet-assisted psychotherapy. *Mims Magazine* (1 December):33–37, 1979.

Corson, S. A., E. O'L. Corson, and P. H. Gwynne: Pet-facilitated psychotherapy. In Anderson, R. S. (Ed.): *Pet Animals and Society*. London, Bailliere Tindall, 1975a, pp. 19–36.

Corson, S. A., E. O'L. Corson, P. H. Gwynne, and L. E. Arnold: Pet facilitated psychotherapy in a hospital setting. In: *Current Psychiatric Therapies*, Vol. 15, J. H. Masserman (ed.). New York: Grune & Stratton, pp. 277–286, 1975b.

Corson, S. A., E. O'L. Corson, P. H. Gwynne and L. E. Arnold. Pet dogs as nonverbal communication links in hospital psychiatry. Comprehensive Psychiatry 18(1): 61–72, 1977.

Darwin, Charles: *The Expression of the Emotions in Man and Animals*. Chicago, University of Chicago Press, 1965.

Davis, Simon J. M. and Francois R. Valla: Evidence for domestication of the dog 12,000 years ago in the Natufian of Israel. *Nature, 276*:608–10, 1978.

Fox, M. W.: *Behaviour of Wolves, Dogs and Related Canids*. New York, Harper & Row, 1971.

Friedmann, E., Thomas, S.A., Noctor, M., and Katcher, A. H.: *Pet Ownership and CHD Patient Survival*. Paper presented at the American Heart Association Meeting in Dallas, Texas, November, 1978.

Gantt, W. H.: Analysis of the effect of person. *Conditional Reflex* 7(2):67–73, 1972.

Harlow, H. F.: *Learning to Love*. New York, Jason Aronson, Inc., 1974.

Hess, E. H.: *The Tell-Tale Eye. How Your Eyes Reveal Hidden Thoughts and Emotions*. New York, Van Nostrand Co., 1975.

Holmes, T. M. and M. Masuda: Life change and illness susceptibility. In Dohrenwend, B. S. and Dohrenwend, B. P. (Eds.): *Stressful Life Events: Their Nature and Effects*. New York, Wiley, 1974, pp. 45–72.

Kane, Robert L. and Kane, Rosalie A.: Care of the aged: old problems in need of new solutions. *Science* 200 (4344): 913–19, 1978.

Levinson, B. M.: *Pet-Oriented Psychotherapy*. Springfield, Thomas, 1969.

_____ : *Pets and Human Development*. Springfield, Thomas, 1972.

Lorenz, K. Z.: *Man Meets Dog*. Baltimore, Penguin Books, 1964.

Lynch, James L., Thomas, Sue A., Paskewitz, David A., Katcher, Aaron H., and Weir, Lourdes O.: Human contact and cardiac arrhythmia in a coronary care unit. *Psychosomatic Medicine* 39(3):188–92, 1977.

Montagu, Ashley: *Touching*. New York, Harper & Row, 1978.

Rahe, R. H.: The pathway between subjects' recent life changes and their near-future illness reports: representative results and methodologic issues. In Dohrenwend, B. S. and Dohrenwend, B. P. (Eds.): *Stressful Life Events: Their Nature and Effects*. New York, Wiley, 1974, pp. 33–86.

Scott, J. P. and Fuller, J. L.: *Genetics and the Social Behavior of the Dog.* Chicago, University of Chicago Press, 1965.

Selye, H.: A syndrome produced by diverse nocuous agents. *Nature* (London) 138:32, 1936.

_____ : The evolution of the stress concept — stress and cardiovascular disease. In Levi, L. (Ed.): *Society, Stress and Disease*, vol. 1. London, Oxford University Press, 1971, pp. 299–311.

Shore, H.: Designing a training program for understanding sensory losses in aging. *Gerontologist* 16(2): 157–65, 1976.

Szasz, K.: *Petishism: Pets and Their People in the Western World.* New York, Holt, Rinehart and Winston, 1968.

Vladeck, Bruce C.: *Unloving Care: The Nursing Home Tragedy.* New York, Basic Books, Inc., 1980.

Yates, E.: *Skeezer: Dog with a Mission.* New York, Harvey House, Inc., 1973.

CHAPTER 9

A CHILD PSYCHIATRIST'S PERSPECTIVE ON CHILDREN
AND THEIR COMPANION ANIMALS

JAMES VANLEEUWEN

ANIMALS AND CHILD DEVELOPMENT

The child psychiatrist is expected to recognize normal child development and perceive the danger signals when things go wrong. In this respect, I wonder about the role companion animals play in the normal development of children. Although others, for example Fox (1976) and Levinson (1972), have written extensively about developmental aspects of the human-companion animal bond, I should like to make a few comments here. Children who never have an opportunity to get involved with animal life would be deprived, I think—deprived of an opportunity that usually gives depth to a wide range of emotional experience. This does not mean that every family should have a companion animal, because there are so many ways in which nature and animal life can be enjoyed. Awareness of the need to provide intelligent care and take responsbility for animals and the ability to feel genuine warmth toward them do not depend on one-time ownership of an animal.

Parents of families who do accept companion animals into their homes should supervise the care of the animals, set boundaries to the intimacy between children and animals, and not permit the animals to infringe upon the freedom of neighbors or the community. The benefit to children of having companions is not derived solely from the animals but depends largely on the parents' psychological awareness of the child's friendship with the animals. No good will come when parents force a child to take

175

responsibility beyond his ability, make the child's friendship with the animal appear ridiculous, or ignore the child's concern when the animal is ill or dies. Normally, parents allow their children to grieve after the death of a companion animal. It might also be a good practice for the veterinarian to help the child express sadness and remove any unnecessary guilt the child may feel because of the death of an animal.

In our hospital there is scarcely a ward without a picture of Charlie Brown, Snoopy the dog, and other members of the well-known gang. Snoopy fits beautifully into the world of children, with their fun, loyalty, and warmth as well as their petty grievances, playful pranks, and serious tears.

Picture in your mind all the fearful children clutching their infinite variety of stuffed animals and think of all the animal books read in a large pediatric hospital and you will once again be impressed by the importance of companion animals.

Given half a chance, every child loves to learn more about nature. Teachers frequently use field trips, nature study, and biology units to boost students' motivation for learning in general. The comment that children learn about sexual behavior and procreation through observation of animals is found like a refrain in almost every talk on companion animals. This is all well and good as long as animal mating behavior is used not as a model for human sexuality but rather as an illustration of variations of behavior, thereby allowing children to broaden their horizons and giving them greater insight into life in general.

From the developmental point of view, there are age related issues that place the companion animal in an ever changing role. During infancy the child has little need for anything apart from the nurturing, "holding" parents. In fact, the infant should be protected from animals and shielded even from their most affectionate ways. Only when the child begins to explore the world can a true interaction with things and animals lead to appropriate relationships. Nevertheless, adult supervision is essential because neither the child nor the animal knows the limits or results of its actions. The animal may not appreciate being dragged around like a toy and may not forever tolerate the manipulations of the "terrible twos."

On the positive side, parents may establish safe conditions within which a companion animal may provide the toddler with stimulation, reality experience, and acceptance. The preschool years present a certain risk when the child asserts himself and experiments with his strength; he can be strong-willed and disobedient. He may blame others and even hand out punishment like a great imitator. Both the animal and the child may become victims during this developmental stage.

I think the greatest benefit from a developmental perspective must be the lively, absorbing experience of play between child and animal. This theme will dominate the relationship for a good number of years of the child's life. For school aged children, a gentle introduction of responsibilities for the care of the animals will emerge smoothly into the twin areas of play and learning alluded to before. And so we could go on to relate the various roles animals play during the continuing life cycle from adolescence to old age.

DISTURBANCES IN CHILDREN'S BEHAVIOR TOWARDS ANIMALS

Family disturbances over children and companion animals fall under three headings in my experience, namely, unfavorable attachments, fears, and cruelty.

Usually, unfavorable attachments are formed when a child has nowhere else to turn for understanding or affection, perhaps in situations of social isolation, loneliness, and depression. Some children with various degrees of disturbance because of emotional deprivation or in some cases childhood schizophrenia show strange preoccupations with flies, ants, birds, etc. One boy I remember behaved, barked, and licked like the family dog.

Keddie's report (1977) on pathological mourning after the death of a domestic pet reminds us of the hazards of pet ownership, where the need for animal companionship has assumed pathological proportions. Understandably the grief upon the death of a beloved pet, which has functioned as a surrogate relative, can go beyond the average expected sadness. He described three patients in pathological mourning who required psychiatric treatment.

Rynearson (1978) placed similar disturbances in the light of

Bowlby's attachment theory (Bowlby, 1977). Under altered circumstances, the normal formation of attachments can go wrong and result in anxious attachments and compulsive care-giving. In the three cases he discussed, the exchange of acceptance and affection between his patients and their pets was less complicated than human exchange of similar needs. This and the preexisting developmentally induced distrust facilitated an intense displacement of attachment to a pet. Rynearson concludes that subsequent separation and loss of such a pet can create complicated psychiatric reactions.

Fear of animals may be very real. There are many situations in which we cannot blame a child for being afraid of cats, dogs, or other animals because he has had a disturbing experience—for example, being bitten. Each year, our hospital treats 300 children for dog bites alone. However, when a child refuses to go to school for fear of seeing even an innocent dog or panics in the presence of a little kitten there is a good chance that the fear is out of proportion to reality and has become a phobia.

There is a spectrum of theories about the origin of such phobias ranging from learning theories such as conditioning, whereby an innocent animal becomes associated with a simultaneous but unrelated frightening stimulus, to dynamic formulations involving otherwise normal mental processes such as symbolization, displacement, denial, and projection. For example, a twelve-year-old girl showed a phobia for dogs and feared kidnappers and the dark. Therapy showed that nobody in the family knew of her hidden fears or anger with her father. She saw him as a secretive, uncommunicative, potentially dangerous man and thought he might actually be a spy. Her fears were displaced and projected into potentially dangerous animals and situations. At the same time, the fears elicited some comfort from the father, who because of them was forced to protect her from external danger. Additionally, her projected fears facilitated her denial of her own angry and otherwise risky impulses by allowing her to perceive the violence she feared as being "out there." Brief psychotherapy with the child and her family restored communication, gave her a sense of reality and perspective, and dispelled her fears.

Disturbances in children may also involve cruelty to animals.

In clinical practice, I find that a certain degree of careless manipulation of small animals is not necessarily a sign of disturbance, particularly when the child is very young or inexperienced. But actions like setting fires to animals, tying the tails of animals together or killing companion animals are sure signs of trouble and warrant psychiatric assessment because such behavior equals desperate cries for help (Hellman and Blackman, 1966; Justice et al., 1974).

One does not have to be a psychiatrist to imagine inner chaos and fear lying at the root of violence. We might just as well turn to the literary talents of William Golding to elucidate the interplay of forces that can eventually lead to inner chaos and the eruption of violence. In *Lord of the Flies*, Golding takes us to an island on which an airplane has crashed, leaving a group of school boys to survive by themselves. Their repeated attempts to create harmony and maintain civilization fail when their childhood frustrations, unbridled fantasy, and fears of abandonment merge with hardships, rivalry, and pride into a spiralling course of fear, anger, hatred, and destruction. Petty arguments are blown out of proportion, and common sense and rituals fail to control the inner fears that come to be perceived as external dangers. The boys start to defend themselves and attack the ghosts and beasts of their imaginations. Scapegoats are found in a relatively defenseless figure—a fat, bespectacled, asthmatic boy—and the small animals of the island. In a climactic development, Golding portrays a particular boy, Jack, moving from psychological inability to kill a pig to the ability to hunt for survival and kill mercilessly to fulfill an inner need. We are made so vividly aware of the complexity of forces determining this action that we can only become more convinced that in the case of cruelty to animals we are not dealing simply with an isolated, impulsive act. In Golding's account, inner fears, real frustrations, threats to survival, past traumatic events, displaced hatred, peer group pressures, chaotic situations, lack of protection, and deprivation of love are all precursors of what might appear to be momentary loss of control; such complex motivation to violence is not restricted to fiction.

Last year I was asked to treat a nine-year-old boy of normal intelligence who had stuffed sand into the mouth of an infant, pinched a piece off his pet rat's tail, maltreated his dog, tried to strangle his cat, put clothes pins on his little brother's fingers, and shaved his guinea pig; otherwise, he appeared as sweet a boy as one could meet. In working with this child, I found a concerned family in agony; this boy had just started to realize that his muscular weakness was the beginning of a slowly progressive illness that would lead to his death in about ten years. He is now reasonably well adjusted. In retrospect, I have no doubt that the unresolved fear and rage about his disability and death, coupled with relatively unsophisticated parental support, produced his transient cruelty as a form of projection and displacement of anger.

CHILD ABUSE, CRUELTY TO ANIMALS, DOG BITES

I have grouped child abuse, cruelty to animals, and dog bites together to bring out the psychosocial similarities and differences involved. Child abuse (defined as a condition of physical harm, malnutrition, emotional neglect, or sexual abuse) has become a major issue of concern in the past two decades.

A hundred years ago animals were apparently better protected against abuse than children. Ironically in 1874, the American Association for the Prevention of Cruelty to Animals took legal action on behalf of an abused child. The girl in this case obtained legal protection on the basis that she belonged to the animal kingdom.

Helfer and Kempe (1968) described "the battered-child syndrome" and ever since, civilized societies have been forced to open their eyes to an unbelievable amount of misery that obviously existed all along but had been conveniently ignored. In one year in New York City, almost 10,000 cases of suspected abuse and neglect were reported (Helfer and Kempe, 1968). Criteria for diagnosing abuse are more firmly established now than in the past. Bruises, broken bones, burns, and malnutrition in young children are no longer blindly accepted as resulting from so-called accidents or diseases without questions of possible child abuse being raised. A high percentage of already abused children will be further abused, severely injured, or killed without intervention.

It would be comforting if child abuse occurred only among

specific groups of people, such as delinquent alcoholics, but this is not the case. Some people are more vulnerable than others, but perhaps we all have a breaking point for losing control.

Corrective punishment is a right, long granted by society to parents. It should be no surprise that some parents carry the "right" to punish their children to the extreme, yet they do this with the best intention of making their children law-abiding citizens. As one parent said to me, "There is so much violence in the world that with my own children I try to nip it in the bud." He severely suppressed all normal assertive behavior in his children.

Many of the abusing parents were themselves physically abused or emotionally deprived as children. They grew up to feel that no matter what they did it was not enough, not right, the wrong time, or in other ways disgraceful. The pattern then repeats itself when such children mature and have children of their own. Without a good model to imitate and without a sense of adequacy or self fulfilment, they often set out to do a better job than their parents did and are overly anxious to make their children behave well so that they will not be criticized as parents. Sometimes they expect their baby to give them the love they missed as children and are disillusioned when the infant appears to have a will of its own.

I feel saddest when I observe parents treating a child as an extension of themselves without regard to the child's own unique and separate being or when the parents' distorted perception makes them think that the infant cries on purpose, makes them blame him for resembling a hated spouse or relative, or in any other innocent way provokes them to feel guilty, inadequate, or angry.

Other factors contributing to child abuse may be found in the child: for example, repeated minor illnesses, miserable temperaments or prematurity, or in the environmental conditions: for example, socioeconomic pressures or shattering disapproval from a spouse or parent, etc.

Veterinarians may not be in the best position to recognize child abuse because they are probably not consulted about such a condition. However, there is an area of overlap between child abuse and cruelty to animals, particularly in families where low frustration tolerance and violence prevail. There are families where everything flies when the going gets rough—dishes, wives, cats, dogs, and children included. It would be sad, therefore, if in analogy to child abuse there persisted a reluctance to recognize the existence of animal abuse among the so-called accidental injuries brought to the veterinarian's attention. Greater awareness of animal abuse may lead veterinarians to initiate mental health intervention for the abusing family in addition to treating the animal.

The similarity in scapegoating of children and companion animals may lead us to suspect that child abuse and animal abuse can be symptoms of the same family disorder. On the other hand, it should be remembered that in many cases child abuse and cruelty to animals are not related because the dynamics are different. For example, the parent may have a distorted perception of the abused child that is uniquely dependent on the parenting role with its multigenerational background and specific interpersonal conflict. A companion animal is not likely to participate in such a specific tragedy. In many cases, therefore, the existence of child abuse in a family does not automatically mean that animals are also abused.

Levinson (1972) implied that the placement of suitable pets might be an aid to children and families where a sense of parental care is lacking, as is the case in child abuse. Without neglecting the potentially positive effects of the human-companion animal relationship, I would not be optimistic about the therapeutic effects of introducing a small animal. Perhaps in some cases there is room for the introduction of a companion animal as an adjunct to an overall treatment plan. It should be remembered, however, that an additional burden of looking after an animal may well tip the balance and accentuate the vulnerable parent's perception of intolerable demands, thus increasing the risk of child abuse.

In my view, child abuse and cruelty to animals should be discussed together with dog bites because they all represent variations on the theme of aggression in the disturbances of psychosocial interaction. In the midst of harm that daily befalls children through accidents, poverty, and war, it may seem trivial to talk about dog bites. But the incidence of dog bites seems to be increasing. The damage inflicted rises sharply with the size of the dog, and currently there is a trend toward having larger dogs as both companion animals and guard dogs. Children are the most frequent victims of dogs and many infants have been severely mutilated or even killed by them.

Our hospital treats 300 victims of dog bites per year. This may well represent a greater amount of physical damage than injuries caused by child abuse. Very little is done to determine the cause and effect of these bites or to find means of preventing them. Attitudes toward dog bites seem to be paradoxical and controversial. People frequently assume that the child victim must be at fault and "had it coming to him." In my experience, this amounts to gross denial of the reality because most children were not provocative; they just wandered unprotected into a situation of being bitten. Blaming children more than the offending animals is simply not valid. There was a time when we similarly held the rape victim more responsible than the rapist. On the other hand, there is no reason to assume that every dog that bites is a vicious animal. Instead of polarizing the issue by blaming either the child or the dog, it seems to me more useful to ask why parents or a particular caretaker or a community allowed the child to be exposed to a hazardous situation.

It is not always easy to decide when responsible parental care ends and negligence begins. One of my patients had her nose bitten off by her grandparents' dog. The child's mother was so severely upset that she needed psychiatric help before she could even look at her own daughter again. Yet the family kept the dog, ignoring further risks to the child. They reasoned that it was a good animal and the four-year-old girl must have put her face too close to it. Dr. Cochrane (1979), a Canadian physician who is well informed on animal-human relationships, states that we should never allow children to eat or carry food in the presence of dogs and we should teach our children not to hold their faces close to animals. If such rules are necessary to prevent children from being bitten, then a lot of people are very negligent.

Indeed, one reason for discussing dog bites together with child abuse is my distinct impression that the mutilation of children by their family dog may result from parental neglect and should be reported to the Children's Aid Society. (The Child Welfare Act of

Ontario, 1978, makes it mandatory that any individual who is aware of child abuse report this knowledge or suspicions to the Children's Aid Society. A physician who does not report suspected child abuse is liable to a fine or imprisonment).

THE THERAPEUTIC USE OF COMPANION ANIMALS

Dr. Levinson (1969, 1972) has not only offered a comprehensive view of the valuable role of companion animals in general but has also explored and described a fascinating variant of therapy that he calls pet oriented child psychotherapy. This technique permits the child to relate to the therapist's dog during the sessions in the office as well as on occasional walks. Dr. Levinson describes the numerous ways in which this contact can be helpful in diagnosis and treatment. For example, it may break the ice for a shy child. Talking to the dog allows indirect communication of problems when direct face to face exploration might be too scary. It may provide an opportunity for an angry, withholding child to show his available affection and spontaneity to the dog. It may provide a reality situation as well as an opportunity for projection of the child's inner feelings. Furthermore, it may allow the child to find reciprocal affection, including physical touch and tenderness, without getting caught up in an exchange of physical affection between therapist and patient.

I believe that from time to time therapists can become very intrigued by a certain technique that in their hands produces exciting results and at which they become more and more proficient. Such expertise is invaluable for thorough understanding and for teaching others the proper use of the technique. It does not always mean that every therapist will be inclined to use such a therapy in his office. I can think of a number of fascinating approaches I have sampled over the years, such as dance therapy, art therapy, behavior therapy, activity group therapy, but I have retained only those with which I feel comfortable within my particular hospital milieu.

From the perspective of having to make recommendations for total care as opposed to just providing office therapy, Dr. Levinson and many others have contributed new ideas for improving living conditions, whether in mental health, correctional, old age or rehabilitation centers. Of particular interest are the papers by Corson et al. (1975) and Mugford and M'Comisky (1975). The introduction of companion animal therapy in residential centers is compatible with the long tradition of other forms of meaningful therapy, such as farming, gardening, occupational, dance, music, and art therapy. It might be useful to differentiate between improvements to the quality of life within an institution, such as the introduction of therapeutic use of companion animals, and a specific form of therapist-patient interaction called psychotherapy.

In preparation for her Master's thesis in the Department of Family Studies, The University of Manitoba, Ferguson (unpublished material, 1979) conducted a survey of the use of pets in residential care facilities for children in Canada. The response from ninety-eight units showed that fifty-four have pets, sixteen have had pets in the past, ten may have them in the future, and eighteen have never had and do not intend to have pets. Further details on the possible therapeutic role of these animals and the interpretation of her findings are not yet available.

Dr. Levinson's writings have, however, made me far more aware of the human-companion animal relationships. Now when I enquire into a child's psychosocial situation, I always include the child's relation to and perception of animals, not only by direct questioning but also through play therapy and picture drawing.

REFLECTIONS ON COMPANION ANIMALS IN PSYCHOTHERAPY

Let me give two case illustrations involving children's relations to and perception of animals.

Carol was an eleven-year-old girl who during the course of psychotherapy frequently expressed the wish for a dog as a companion. This seemed a reasonable request considering her shy but pleasant nature, lack of social skills, and tendency to watch television as her favourite pastime. She lived in a quiet, safe neighborhood in a sizeable home with a fenced back yard. She was an only child and her father had died a year before I saw her, increasing her loneliness to the level of a chronic depression.

Anyone would love to see such a child experience the joy a companion animal can bring to a lonely person.

On the other hand, Carol and her mother had come for help because of the conflict in their relationship that was hidden from the outside world. Despite her shy nature, Carol controlled her mother at home by loud, abusive language, negative behavior, and destructive temper tantrums thrown upon minimal frustrations. I learned that the mother had always been an ineffectual parent, and her prolonged, extreme grief after the death of her husband had not made her any stronger in discipline or more available to the child's needs. The father had spoiled his little girl. Carol erroneously believed she had caused her father's death and thought she was bad and deserved punishment. She had been unable to grieve properly.

Within this context and after some time in therapy, the mother asked me to consider the wisdom of acquiring a dog for Carol. They had already made considerable progress in their relationship, but I suggested they postpone their decision to select a dog until Carol managed to take greater personal responsibility in certain daily routines and the mother had established stronger parental control. Furthermore, I insisted on Carol and her mother reading about pet care and reaching an agreement on areas of responsibility in caring for a dog. They also accepted my advice to spend time with friends of mine who have a delightful way with dogs. Finally, they acquired a dog but gave it up after a few months because they could not train it properly. (Now I might consult a book such as William E. Campbell's *Behaviour Problems in Dogs*, 1975, or be more assertive in referring them to an animal behavior therapist.)

They acquired a second dog. Fortunately this time, the veterinarian offered considerable assistance and they did very well. One day this dog escaped from the house and was run over by a car. Carol grieved profoundly. Through the experience, she seemed able to allow the much deeper pain caused by her father's death to emerge and to be shared with her mother and her therapist. This ability to share and master depressive feelings was helpful in Carol's resolution of her frequently masked depression.

Louise grew up in an unsettled home environment. Her older

brother had a psychotic illness in childhood that made him have strange preoccupations and an autistic inability to communicate with people. Another brother was very temperamental and always beset by fears. The parents were frequently at their wits end. The mother had gone through a serious depression, and the father had become authoritarian despite his very sensitive nature. At one point, the parents asked me for help in managing Louise's rambunctious, hyperactive, and childishly silly behavior. To my surprise, this seven-year-old girl decided to act like a friendly animal during the initial phases of therapy. Her crawling, diving, gnawing at chairs, wiggling her behind as though trying to flap her tail, and making of various sounds showed that she was a beaver. Slowly she allowed me access to her imaginary beaver family.

This bright, creative youngster had fabricated a delightful world of playful sibs and marvellously caring beaver parents. She had read most children's nature books on beavers; her teacher considered her an expert in the field and allowed her to instruct the class on beaver behavior. However, her intense identification was shared only with her therapist because she trusted that I would not laugh at her or tell anyone. The fantasy world had become the escape she fled to in the privacy of her bedroom. When reality became too difficult, she found a refueling station in the creative act of staging a life away from life.

This girl needed help to make sense of the disturbance in her family. In a playful yet serious, empathetic relationship, we were able to compare and contrast her real and imagined family life. Both family therapy and individual psychotherapy served to foster appropriately friendly relationships and allowed her to accept the strengths and weaknesses of her family members.

In two years, at age nine, when her condition was much improved and she was socially no longer in difficulty, I asked her to draw a picture of all the members of her family doing something (Fig. 9-1). The result was fascinating. The size of her family members did not relate to reality, and there was little togetherness. The two brothers were still disliked; they were called "dumb" and symbolically crossed out. The companion animals received a very prominent position in her perception of the family. Not even a lay person could resist wondering about the significance of the urinating dog. It took us via the usual "pee and pooh" phase to discuss some related areas of psychosexual concern.

Figure 9-1.   Family drawing made by a nine-year-old girl.

ANIMALS IN CHILDREN'S DRAWINGS

The "Draw A Person" test has been used for fifty years to estimate a child's intelligence (Goodenough, 1926). The scoring is done by adding up points given for including parts, i.e. head, arms, feet, fingers, etc. Later the "House-Tree-Person" test added the possibility of learning more about the personality of the child by examining the reflected sensitivity, flexibility, vitality etc. (Buck, 1948). In the "Draw A Family" test the child may express certain personal or family conflicts (Hulse, 1951).

Burns and Kaufman (1970, 1972) introduced an interesting new approach to the diagnostic use of children's drawings. Rather than the akinetic instructions that usually result in relatively static, rigid drawings, they instructed the children to use action in their drawings by asking the child to draw everyone in his family doing something. Themes of aggression, isolation, and danger are easily recognized in contrast to the family atmosphere of closeness, warmth, and confidence that is normally expected.

Drawing people performing actions is a simple attractive task that most children find fun. It could be used as a projective test like the Rorschach ink blot test or other projective techniques. Burns and Kaufman offer guidelines in their interpretative manual called *Actions, Styles and Symbols in Kinetic Family Drawing* (1972). They give the example of the vacuum cleaner that is a powerful, controlling, cleaning device and point out that it may be significant that some children draw their mothers vacuuming, while others portray their mothers in a nurturing role. A vacuum cleaning mother in the drawing may symbolically alert us to the child's perception of the mother as an overpowering, controlling person.

I have found it interesting to give the Kinetic-Family-Drawing task to most children I see. The drawings do not provide a short-cut to diagnostic assessment, but they do offer confirmation of elaborations of certain themes and often facilitate discussion of the family situation.

The following examples are drawings by thirteen-year-old children.

Figure 9-2 is by a normal boy who drew his family on vacation together. Figure 9-3 was drawn by a boy from a normal control group. When this drawing was made, I expressed some concern about the father's precarious position of the slope. To my surprise, I discovered in my follow-up that the father deserted the family several months after the drawing was done. Figure 9-4 was done by a boy who had been disturbed for many years. He was recovering from childhood schizophrenia. Despite his normal intelligence, he drew a primitive, compartmentalized picture of his family. Figure 9-5 is by an insecure, shy girl who is especially attached to

Figure 9-2.   Family drawing made by a healthy thirteen-year-old boy.

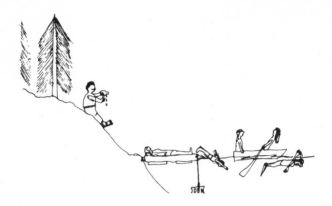

Figure 9-3.   Family drawing made by a healthy thirteen-year-old boy.

Figure 9-4.   Family drawing made by a disturbed thirteen-year-old boy.

her dog. Figure 9-6 is by a thirteen-year-old delinquent boy from an
antisocial family with alcoholism, deprivation, and violence. The
picture reflects rigidity and a strict pecking order. The dog that is
being fed was seen by this boy as a preferred member of the family.
The senior, most aggressive members of the family have no hands.

In their analysis of many pictures, Burns and Kaufman (1972) came to the conclusion that preoccupation with cats is often symbolic of conflict in identification with the mother. They felt that the furry "cuddliness" of the cat combined with its teeth and claws, created a symbol related to ambivalence and conflict. However, they also show that cats can be associated with love, especially love transferred from the parents. Like any other symbol, the cat can

Figure 9-5. Family drawing made by an overly shy thirteen-year-old girl.

Figure 9-6. Family drawing made by an antisocial thirteen-year-old boy.

have various meanings. Figure 9-7, drawn by an eight-year-old, combines cats with the family at the park.

Figure 9-7.   Family drawing made by an anxious eight-year-old girl.

According to the theory of Burns and Kaufman, the person at the top of the ladder might be associated with tension and precarious balance. The prominence of the two cats might call our attention to questions of warmth, love, and security. The picture does reflect a sense of happy togetherness, and in therapy I could confirm the existence of a genuine attempt to make life pleasant. The girl was not sure whether to accept the mother's new boyfriend who had recently moved into their house. It is this potential father figure who is drawn out of proportion to his size, performing the balancing act at the top of the ladder. This girl was brought to therapy because of her anticipatory fear of accidents happening to her family and her simultaneous upsurge of clinging behavior. Although the cats were not really taken to the park, they are shown in the picture because they meet a certain need for warmth, love, and security in both symbolic and actual terms. The young patient explained that the cats are of course not members of her family and do not mean as much to her as people do, but the companionship makes her feel good and less afraid.

Burns and Kaufman also mention the symbolism of the horse and the snake appearing in their collection of drawings, but there is no comment on the dogs. Yet I counted dogs in 15 percent of their illustrations. I have always found it curious that a percentage of children should include their companion animals in a drawing that specifically deals with the members of their family doing something. In my experience, most children who have companion animals do not draw them because they know animals are not part of their family. Of those who draw animals in this context, most comment that they are part of their environment or part of the action without elevating the animal companionship to the level of human relationships. However, I am no longer surprised to hear a number of children explain that their animals appear in their pictures because they seem more like family members than either the parents of the sibs or the whole family put together. The child's perception is not always in keeping with the reality, but many such children are truly unloved, have real behavior problems, and are socially inept or lonely.

CONCLUSION

The interest is companion animals is evident in publications such as *Do We Need Dogs* (Adell-Bath et al., 1976) from Sweden, *The Dog Crisis* (Nowell, 1978) from Canada, *Pets as a Social Phenomenon* (1976) from Australia, and this book from Great Britain and America. There is still some tension between those who enthusiastically promote liberal distribution of companion animals and those who caution against overpopulation and health hazards and are concerned about the many other problems related to domestic animals.

We need a balanced perspective of the role of companion animals in the lives of children. Through an interdisciplinary approach combining the social, medical, and veterinary sciences, we can achieve this.

REFERENCES

Adell-Bath, M., Krook, A. C., Sandqvist, G., and Skantze, K.: *Do We Need Dogs? A Study of Dog's Social Significance to Man.* Unpublished thesis for the University of Gothenburg, Gothenburg, Sweden, 1976.

Bowlby, J.: The making and breaking of affectional bonds. *Brit J Psychiat, 130*: 207, 1977.

Buck, J. N.: A qualitative and quantitative scoring manual. *J Clin Psychol, 4*: 317, 1948.

Burns, R. C. and Kaufman, S. H.: *Kinetic Family Drawings (K-F-D). An Introduction to Understanding Children through Kinetic Drawings.* New York, Brunner/Mazel, 1970.

———: *Actions, Styles and Symbols in Kinetic Family Drawing (K-F-D). An Interpretative Manual.* New York, Brunner/Mazel, 1972.

Campbell, W. E.: *Behaviour Problems in Dogs.* Santa Barbara, American Veterinary Publications, Inc., 1975.

The Child Welfare Act, 1978. S. O. 1978.

Cochrane, B. M.: *Your Pets Your Health and the Law.* Toronto, John Wiley & Sons, New York, 1979.

Corson, S. A., Corson, E. O., and Gwynne, P. H.: Pet-facilitated psychotherapy. In *Pet Animals and Society: A B.S.A.V.A. Symposium.* London, Bailliere Tindall, 1975.

Ferguson, Personal Communication, 1979.

Fox, M. W.: The needs of people for pets. In *Proceedings of the First Canadian Symposium on Pets and Society, and Emerging Municipal Issue.* Ottawa, Canadian Federation of Humane Societies and Canadian Veterinary Medical Association, 1976.

Goodenough, F. L.: *Measurement of Intelligence by Drawings.* New York, Harcourt, Brace and World, Inc., 1926.

Helfer, R. E. and Kempe, C. H. (Eds.): *The Battered Child.* Chicago, The Univer- of Chicago Press, 1968.

Hellman, D. S. and Blackman, N.: Enuresis, firesetting and cruelty to animals: a triad predictive of adult crime. *Am J Psychiat, 122*: 1431, 1966.

Hulse, W. C.: The emotionally disturbed child draws his family. *Quart J Child Behav, 3*: 152, 1951.

Justice, B., Justice, R., and Kraft, I. A.: Early-warning signs of violence: is a triad enough? *Am J Psychiat, 131*: 457, 1974.

Keddie, K. M. G.: Pathological mourning after the death of a domestic pet. *Brit J. Psychiat*, 131: 21, 1977.

Levinson, B. M.: *Pet-Oriented Child Psychotherapy.* Springfield, Thomas, 1972.

Levinson, B. M.: *Pets and Human Development.* Springfield, Thomas, 1972.

Mugford, R. A. and M'Comisky, J. G.: Some recent work on the psycho-therapeutic value of cage birds with old people. In Anderson, R. S. (Ed.): *Pet Animals and Society: A B.S.A.V.A. Symposium.* London, Balliere Tindall, 1975.

Nowell, I.: *The Dog Crisis.* Toronto, McClelland & Stewart, 1978.

Pets as a Social Phenomenon. A Study of Man-Pet Interactions in Urban Communities. Melbourne, Petcare Information and Advisory Service, 1976.

Rynearson, E. K.: Humans and pets and attachment. *Brit J Psychiat, 133*: 550, 1978.

CHAPTER 10

THE PET DOG IN THE HOME: A STUDY OF INTERACTIONS*

ALASDAIR MACDONALD

THE HOUSEHOLD PET

Both Levinson (1969, 1972) and Bridger (1970, 1976) have suggested that contact with pets is generally important in human development. Their opinions were based not on systematic studies but on long experience of work with abnormal individuals. The authors, however, were not able to present data from normal subjects for comparison. From his very detailed studies of a small number of abnormal families, Speck (1964, 1965) suggested that pets have a role to play in communication within abnormal families, but again, no data were available from normal subjects.

While working in the child psychiatry clinic at the University of Dundee Department of Psychiatry, I encountered a case of cruelty to a pet animal. Upon consulting the literature for background information I found that little was available either about cruelty to animals or about the normal relationship between children and their pets. I decided to investigate this field and to examine the normal pet-owning family first of all.

Dogs were chosen for the study because they are the most common household pet in the United Kingdom (Anderson, 1975) and because they offer the greatest variety of interactions. One parent and one child from each family was asked to complete a questionnaire as it was thought that this would be sufficient to give an outline of the relevant interactions.

The questionnaire was designed partly with structured responses

*Thanks are due to Dundee District Department of Education for its cooperation and to Dr. C. V. Gamsu for advice.

195

to simplify answering and data processing and partly with open-ended questions where it was undesirable to limit the range of possible answers.

The questionnaire was completed by a number of volunteer parent-child pairs before the actual study began, and they reported no difficulties. It took about twenty minutes to complete each questionnaire.

Near random samples of ten-year-old children were selected by choosing one class from each of six schools, five being urban and one rural. These schools were selected from areas of varied social background. Children ten years of age were selected because at that age family centered activity would still predominate, but the child would be old enough to adequately answer the questions.

All the children with pet dogs were given both a questionnaire and a letter for their parents explaining the study and asking their cooperation. Those whose parents returned the completed questionnaire then completed one themselves, unassisted. The parent's questionnaire had forty-eight items, and the children's questionnaire had forty-three items.

There were thirty-one pairs of completed questionnaires returned. The low response rate was disappointing in view of the attempt made to enlist the interest of the individual children, and it was thought that two factors might be responsible. The number of dog licences issued is known to be vastly exceeded by the number of dogs in the community, and owners of unlicensed dogs may have feared repercussions. Also some local authority housing is governed by regulations forbidding the keeping of dogs on pain of eviction, which would provide a strong motivation for keeping the dog out of the public eye as much as possible. The two schools in lower social class areas had the lowest response rate, and it might be in these areas that these factors would be strongest.

DEMOGRAPHIC DATA

Social class (by occupation of head of household) is shown against the Dundee (1971) census figures in Table 10-I.

The bias toward the upper social classes is not uncommon in studies using self-report questionnaires.

TABLE 10-I

SOCIAL CLASS COMPARISON

SOCIAL CLASS	I	II	III	IV	V
Dundee (1971) %	4	12	50	22	12
This study %	13	19	45	13	3

There were eighteen girls and fourteen boys. Mean age was 10.68 years (S.D. $\pm$ 0.48). There were sixteen children who had older siblings only, and seven had younger siblings only. One child had no siblings. In the subsequent results, no effect was found for sex, age, birth order, or school attended.

PARENTS' ATTITUDE

Parents' reasons for the choice of a particular dog are given in Table 10-II.

TABLE 10-II

PARENTS' REASONS FOR CHOICE OF DOG

Behavioral attributes	19
"We like all dogs"	18
Suitable size	15
Specific breed selected	14
Special abilities	10
Suitable as watchdog	10
"Rescued"	6

The "rescued" category represents those cases where the dog had originally belonged to a friend or relative and would have been destroyed if not taken in by its present owners. More than one reason was given for the choice of dog in many cases, e.g. those who "liked all dogs" also mentioned behavior as a factor in their choice in nine cases.

Some potential problems of dog ownership had been considered by parents beforehand and are listed in Table 10-III. Concern over the potential problems following loss of the dog was most prominent.

TABLE 10-III

POTENTIAL PROBLEMS OF DOG OWNERSHIP

Effect of loss on child or family	23
Responsibility of care	22
Danger of dog annoying neighbors, etc.	14
Danger of injury to child by dog	12
Danger of child illtreating dog	10
Danger of child catching disease from dog	10
Damage to others' property by dog	9

Contrasting with these potential difficulties are the parents' comments about the benefit of dog ownership. The most frequent responses were that the child would "learn responsibility" (15) and would learn to "love" or "respect" animals (13).

Levinson (1969, 1972) has suggested that children can usefully learn about friendship, toilet training, sexual behavior, and pregnancy from pet dogs. Parents were asked specifically if this had occurred for each area of activity and whether they thought this a relevant factor in keeping a dog. There were twenty-one parents who thought that their child had learned about friendship from the dog, and six who felt they had learned about toilet training.

All thought this worthwhile. There were thirteen parents who thought that their child had learned about sexual behavior and eleven who thought that this was valuable, although only eight were in both groups. Similarly, nine parents thought that their child had learned about pregnancy, and ten thought this beneficial, but only seven were in both groups. Only two dogs in the study had had puppies, so the latter responses probably represent opinions rather than observed facts.

Levinson expected that preschool children would benefit most greatly from pets. However, only thirteen parents in this study expected any benefit to a preschool child, the emphasis falling more in the late primary school age group. This was borne out by the finding that eighteen of the children had been over eight years old when the dog was obtained and that only two parents thought that the dog could usefully have been obtained sooner. C. Perin's theory (*see* Chapter 4 this book) is reinforced by this finding.

Looking after a pet is also believed to teach children self-restraint. There were twenty-two parents who thought this was true, but only five volunteered this in answer to open-ended questions, implying that this was not of great significance to them.

These findings suggest that many parents consider dog ownership and its effects with some care. Nevertheless, only ten parents thought that their child would have suffered by not having a dog.

CHILDREN'S ATTITUDES

The dog was described as belonging to the child in seven cases and to the whole family in twenty-one, matching closely the parents' replies to the same question. There were twenty children who thought that the dog was intended to be a companion and six who proposed it as an object for giving or receiving affection. There were other pets in the household in fourteen cases. The dog slept in the child's room in thirteen cases. A career involving work with animals was envisaged by twenty-two children. There were twenty-six children who played more with the dog than did any other family member, and seven reported frequent play involving physical contact with the dog, e.g. "wrestling." The majority of dogs were fed by the mother even when the child had been described as being in charge of all care.

The dog was exercised mainly or exclusively by the child in twenty-three cases. Social contact with other children or adults while exercising the dog was reported by twenty-six children, supporting Levinson's (1969) suggestion that a lonely child can increase his social contacts while exercising the dog locally.

There were seventeen children who talked more with the dog than did any other family member. There were seven parents and eight children, from the same families in four cases, who often talked to the dog about general topics unrelated to the dog itself, e.g. other family members: "about my Dad." The dog was believed by twenty children to understand all their speech, by twenty-two to respond to their mood, and by twenty-five to understand all their feelings, "He knows when I am sad."

HIGH INTERACTION RESPONSE

A number of responses were seen to occur frequently. The relevant categories were that the child took a major part (1) in play with the dog, (2) in exercising the dog, and (3) in talking to the dog. Responses also made frequently were those concerning the dog's perceived understanding of feelings, mood, and speech. These six categories were grouped together as "high interaction categories" and were studied in more detail. The responses in the six categories for each case are tabulated in Table 10-IV.

Many of the children interact highly with the dog and at the same time perceive the dog as understanding the content of their communications. Only a small group see the dog as understanding but do not interact with it. It was not possible to identify a reason for this, although one or two of these children indicated that another family member spent much time with the dog, perhaps thus excluding the child.

In the parent questionnaire, eighteen children were reported to look after the dog without help. Ten of these children fitted into all the previously mentioned high interaction categories.

There were a number of responses, mostly to open-ended questions, that seemed likely to indicate a close relationship between child and dog. Individually these responses were not frequent,

TABLE 10-IV

RESPONSES IN HIGH INTERACTION CATEGORIES FOR ALL CASES

No.	Child plays with dog	Child walks dog	Child talks to dog	Dog knows feelings	Dog knows mood	Dog knows speech
1	+	+	+	+	+	+
2	+	+	+	+	+	+
3	+	+	+	+	+	+
4	+	+	+	+	+	+
5	+	+	+	+	+	+
6	+	+	+	+	+	+
7	+	+	+	+	+	+
8	+	+	+	+	+	+
9	+	+	+	+	+	+
10	+	+	+	+	+	+
11	+	+	+	+	+	
12	+	+	+	+	+	
13	+		+	+	+	+
14	+	+	+	+		+
15	+	+		+	+	+
16	+		+	+	+	
17	+	+	+	+		
18	+	+		+	+	
19	+	+		+	+	
20	+		+	+		
21	+	+				+
22	+			+	+	
23	+					+
24	+					+
25	+					
26	+			+	+	+
27		+				+
28		+		+	+	
29		+		+	+	
30						+
31		+		+	+	+
	26	23	17	25	22	20

child and dog. Individually these responses were not frequent, but they were examined in more detail to identify possible points of interest for further investigation.

Of the eight children who talked to their dog about general topics, five were included in all six high interaction categories, and only one was not in at least three. The parents identified six as looking after the dog without assistance.

Play involving physical contact was reported by seven children, and of these four were in all the high interaction categories. Again only one child was not in at least three. There were five among the seven children who regarded the dog as belonging to them, and four among the six children who saw the dog as an object of affection. All seven were reported to look after the dog without help.

There were seven children who regarded the dog as belonging to them, and of these four were included in all the high interaction categories and only one was in less than four. Parents reported that four of these children looked after the dog themselves.

The category "Dog seen as an object for giving or receiving affection" was answered for six children of whom four were in all the high interaction categories. Parents reported that half of these children looked after the dog without help.

On the basis of the results available, it would seem that if a child is given a dog for him or herself and looks after it without help, then there is likely to be a high level of interaction between them. A selection process may be at work here because the child's attitude to dogs may have affected the parental decision to obtain the dog originally. The overlap between the high interaction categories and the various smaller categories suggests that the interaction may be quite intense emotionally, whatever its origin.

THE CLINIC GROUP

To obtain a clinical context, the same questionnaires were used with a child psychiatric group. New referrals to an urban child psychiatric clinic were questioned at their first appointment.

Seven completed pairs of questionnaires were obtained. The clinic sample did not differ appreciably from the main sample in terms of age, sex, or birth order, and social class data were similar.

Information from parents was more limited in the clinic sample, perhaps due to the setting. It was noted that the dog had been obtained for the child by five of the seven parents in this group, as compared to six of the thirty-one parents in the main sample.

When looking at parents' responses for factors considered before obtaining a dog, the risk of being upset by its loss headed the list again, being considered important by five parents. However, the responsibility of owning a dog was considered by only three parents and the risk of damage to others' property by only one (as compared to twenty-two and nine respectively in the main sample).

In the children's responses, a somewhat similar pattern to that of the main sample emerged in relation to the high interaction categories (*see* Table 10-V). As regards the smaller categories, three of the children regarded the dog as belonging to them and two reported play involving physical contact. No child or parent was reported as talking to the dog about general topics, and no child identified the dog as an object for giving or receiving affection. Social contacts while exercising the dog were reported

TABLE 10-V

RESPONSES IN HIGH INTERACTION CATEGORIES FOR CLINIC CASES

No.	Child plays with dog	Child walks dog	Child talks to dog	Dog knows feelings	Dog knows mood	Dog knows speech
1	+	+	+	+	+	+
2	+	+	+	+	+	
3	+		+	+	+	+
4	+	+	+			+
5	+	+			+	
6				+		
7		+		+		
	5	5	4	5	4	3

by six of the children. Parents reported that three children looked after the dog themselves.

With only seven pairs tested, the numbers are too small for true comparison or statistical analysis, but some differences between the two groups are interesting.

In the parent enquiry, having a pet dog "to prevent fear of dogs" in the child was quoted in response to open-ended questioning. This carries implications in respect of the parents' attitudes to the child's expressed or potential fears. There were three parents in each sample who made this comment, which may indicate a difference between the attitudes of these two groups but was not statistically significant.

In most respects, the main sample and the clinic sample made similar responses. However, the clinic parents had more often obtained the dog for the child and as mentioned had often considered the child's attitude to dogs when doing so. The clinic parents' responses suggested less concern for the possible effects of the dog on persons outside the family.

There were structured and open-ended questions about cruelty included in both questionnaires, but responses were minimal. This may be due to socially acceptable responses being favored in this type of questionnaire. The possibility of the child illtreating the dog was considered by ten parents in the main sample and by three in the clinic sample before obtaining the dog.

THE DOG AS A "SIGNIFICANT OTHER"

The most prolific writer in the field of human-pet interaction has been Levinson (1969, 1972). For example, he believes that a dog or other pet can act as a confidant and emotional support for an abnormal child. The findings of this study suggest that some of his intuitive proposals may also be true for normal children. The findings for the high interaction categories in this study suggest that normal children who interact closely with the dog also believe the dog to have an understanding of the emotional content of the interaction. Those children who talk to the dog about general topics may be using it as a confidant, but there is no direct evidence for this. Aaron Katcher (*see* Chapter 3) has more to say on this subject. Children without siblings might be a fruitful group for further study in this context, but there was only one such child in the present study.

Levinson's work indicates that the pet may function as a "significant other" for the child in attachment theory terms. The concept of attachment in relation to normal human-pet relationships has been ably discussed by Rynearson (1978). He says, "....human and pet are significant attachment figures for one another. Under normal circumstances they share complementary attachment because of mutual need and response. At times of stress they may temporarily seek out the other for attachment."

In other words, the relationship between pet and human shares many of the features that identify important bonds between humans, such as mother-child relationships. It seems likely that those children with several entries in the interaction categories have a significant attachment to their dogs and that the relevant items from the questionnaire could be used to identify such children.

Distress occasioned by separation or loss is important in relation to attachment. In the present study, the possibility of being distressed by the loss of the dog was mentioned by twenty-three parents, and it was the factor most frequently considered before obtaining a dog. Of the seven clinic parents, five also thought this important. This may represent opinions formed in retrospect once having become attached to the particular dog or may be due to having experienced the loss of previous dogs. However, it implies an acceptance by the general public of distress at the loss of a dog and suggests that degrees of mourning for dogs may be more widespread in the community than has been thought (Macdonald, 1977; Keddie, 1977). If this is correct, then it is relevant to veterinary practice. If the loss of a dog is such an important event that it is thought of even before the dog is in contact with the family, then the eventual death of the animal, whether natural or assisted, may have considerable emotional impact.

The low response rate in this study, discussed previously, means that the results must be interpreted with caution. Nonetheless, it appears that dogs may function as objects of attachment for at least some normal children, and this has implications for child care, for veterinary practice, and for these disciplines' effect on one another.

REFERENCES

Anderson, R. S. (Ed.): *Pet Animals and Society*, Bailliere Tindall, London, 1975, Chapter 5.

Bridger, H.: Companionship with humans. *Proc. 77th Congress Roy. Soc. Hlth.*, pp. 1970, 166–70.

_____ : The changing role of pets in society. *J Small Anim Pract, 17*: 1–8, 1976.

Keddie, K. M. G.: Pathological Mourning after death of a domestic pet. *Brit J Psychiatry, 131*: 21–25, 1977.

Levinson, B.: *Pet Oriented Child Psychotherapy*. Thomas, Springfield, 1969.

_____ : *Pets and Human Development*. Thomas, Springfield, 1972.

Macdonald, A. J.: Mourning after pets. *Brit J Psychiatry, 131*: 551, 1977.

Rynearson, E. K.: Humans and pets and attachment. *Brit J Psychiatry, 133*: 55–55, 1978.

Speck, R. V.: Mental health problems involving the family, the pet and the veterinarian. *J Am Vet Med Ass, 145*: 150–54, 1964.

_____ : In Friedman, A. S. et al. *Some Specific Therapeutic Techniques with Schizophrenic Families. Psychotherapy for the Whole Family*, Springer Publishing Co., New York, 1965.

SECTION III

THE HUMAN–COMPANION ANIMAL BOND IN SOCIAL PLANNING AND IN VETERINARY SCHOOLS

CHAPTER 11

ESTABLISHING A SOCIAL WORK SERVICE IN A VETERINARY HOSPITAL

ELEANOR L. RYDER AND MARIALISA ROMASCO

T he Companion Animal Clinic in the Small Animal Hospital of the University of Pennsylvania's School of Veterinary Medicine was established in the fall of 1978 as a multidisciplinary effort to develop procedures for dealing with some of the people — animal problems that surface in a veterinary hospital. From this experience we hoped to develop teaching materials for veterinary students and for human services personnel. (The phrase "human services personnel" is used here to refer to psychiatrists, psychologists, social workers, and other persons who may be involved in the delivery of community mental health services.)

HISTORY, PURPOSE STRUCTURE

Organizational structure has played an important part in the development of the work at the University of Pennsylvania. There are three organizational arrangements staffed and financed by two separate but related projects.

The first of these, the one in which I work, is a federally funded training project entitled *Veterinary Medicine and the Study of Human Behaviour.* The second, directed by Dr. Alan Beck, is the *Centre for Interaction of Animals and Society*, a research and training program funded by private contributions. The third is the Companion Animal Clinic, an operational arm of the *Veterinary Medicine and Human Behaviour* project.

Some knowledge of how and why the project developed will, I think, help in the understanding of the significance of the clinic.

Veterinary Medicine and Human Behaviour, funded by the United States Public Health Service, was designed by Dr. Aaron Katcher,

209

a psychiatrist interested in studying the nature and significance of emotional ties between people and their pets. He requested funds for research as well as for training. He hoped that if the Public Health Service would support the research of a multidisciplinary team to study the human–companion animal bond, data could be generated for use in teaching veterinarians and human services professionals. The impetus for the project came from a growing realization at the University of Pennsylvania that veterinary students should have better preparation for understanding and dealing with or referring the human problems that come to them and that human service professionals need to understand the significance of the role of pets and their veterinary physicians in the health—physical and mental—of human beings.

The funding that was provided by Public Health Service was for the support of the training aspect of the proposal but not for research. We therefore were in the position of having to develop a research program out of scarce university resources.

In the meantime, the Dean of the School of Veterinary Medicine had approached the Dean of the School of Social Work to discuss the feasibility of developing a specialty of social work in veterinary medicine similar to that of social work in human medicine. The funding of the training program provided an opportunity to test this idea with project trainees. It was out of this combination of interests and resources that the Companion Animal Clinic was formed. Dr. Washington, our first postdoctoral veterinarian, joined the project staff in July of 1978, and Miss Romasco, a predoctoral social work student, began in September. Other staff of the Clinic included three faculty members: Dr. Katcher from the Department of Psychiatry, Dr. Sheldon Steinberg from the School of Veterinary Medicine, and myself from the School of Social Work. Our Administrative Assistant, Miss Debrah Meislich, doubled as a research assistant as did several undergraduate students.

In preparation for opening the clinic, staff members spent several weeks observing veterinary hospital procedure and discussing plans for the service with each other and with veterinarians in the small animal hospital. In mid-October we were invited to present plans for the Companion Animal Clinic to the hospital staff. We developed the following statement of our expectations:

"A major purpose of the Companion Animal Clinic is to begin to develop a body of knowledge about the interaction of people and animals, and their effects on each other. Clinic staff would like to know about any 'people problems' you may encounter, to learn how you deal with them and, if you are willing, to talk with the people involved. In return, we expect to provide some help to people with pet problems and to the treating veterinarians who must deal with them, whose work may be impeded by pet owners."

Along with this statement of purpose, staff had listed four kinds of problems with which we thought the Companion Animal Clinic might be able to give specific help:

1. Emotional problems of pet owners whose pets are seriously ill or have died.

2. Animal problems that may reflect individual or family problems of owners.

3. Problems with owners who are unreasonable or irrational.

4. Problems with people who want some help in selecting the right kind of pet for them or their family.

It was obvious from discussion that some veterinarians were reluctant to engage in the project with a psychiatrist and social workers. Several expressed their concern that what we were proposing might interfere with the relationship between veterinarian and owner and somehow lessen its importance. Our assurance that we hoped to be able to enhance that relationship did not seem to reduce their anxiety.

For the social work staff, there were two factors that reassured us on this matter. One was that social workers are accustomed to operating in a host setting, and part of our professional skill lies in our ability to carry our responsibility without interfering with professional-client relationship in the supporting service. The second factor was related to the nature of the setting. The Small Animal Hospital at the University of Pennsylvania is a teaching hospital rather than a primary care institution. Most patients are referred by veterinarians in the surrounding community. This means that veterinarians in our hospital rarely have ongoing professional relationships with the owners they see, a point of significance when considering the transferability of a social work service to a primary care institution.

In addition to hospital staff reluctance, we faced a serious space problem when the clinic finally opened late in October. A new building is under construction which, when completed, will approximately double the space for the hospital; in the meantime, there literally was no corner that could be designated solely for our use. We were given access at specified times to a large examining room that also housed an autoclave, so there was little guarantee of privacy for conferences with clients. The room had no telephone or desk. Dr. Washington and Miss Romasco shared a small cubicle with a veterinary resident, using his desk, chair, and telephone as they could. Faculty were housed in three different buildings. In spite of all of this the clinic opened.

From the outset, it was agreed that whoever was available should see the owner and take the initial information to be recorded on an intake form we had developed. The staff would meet each week to review cases and decide how and by whom they should be handled. It was anticipated that Miss Romasco would handle problems that were primarily those of the people involved, unless there was evidence of severe emotional disturbance. These cases were referred to Dr. Katcher. Problems of animal behavior and control could be identified and handled by Dr. Washington.

CLINIC BEGINNINGS

Many of the veterinarians, focused as they were on the problems of sick animals, were not sure that a social worker could give them any help or, if she could, that they really wanted it. The first cases that they referred to Miss Romasco were primarily nuisance cases—people who made excessive or irrational demands or accused the veterinarian of incompetence.

An early case was that of a woman whose pet had died in the hospital and who was threatening to sue the veterinarian and/or the hospital. Miss Romasco was able to help her deal with her anger and grief and, after several sessions, to come to the conclusion that the hospital staff had done all it could for the pet. When she offered a financial gift as an expression of her appreciation, several of the veterinarians decided that perhaps we could be helpful by taking some of the less pleasant humans out of their offices and off their telephones.

We were pleased with this success, but this kind of case did not really give us an opportunity to explore the interdisciplinary potentials of two professions working together. We needed an opportunity not only to offer a social work service to clients but also to work with veterinarians on assessing people–animal relationships in families where pets are maintained primarily for companionship.

Fortunately for the clinic, the chief of oncology, Dr. Robert Brodey, was interested in learning whether we could, in fact, help him and his students work more effectively with owners whose pets were terminally ill—people who had to make choices between no treatment, expensive treatment, and/or euthanasia. He invited Miss Romasco to participate in his clinic rounds, and with his guidance she developed procedures for working in other services as well as oncology.

Clinic procedure in the hospital at the University of Pennsylvania begins when veterinary students meet the pet owner, take a history of the problem, and make a preliminary examination of the animal. One student does this while other students assigned to that service watch. There may or may not be discussion among the students, and if there is it may or may not involve the owner. When the examination is finished, the student who had made it goes out of the room to report to the senior veterinarian on the observation, diagnosis, and recommended treatment. The senior veterinarian then goes to the examining room with the student and repeats any aspect of the examination and history taking that he feels is necessary in order to make his diagnosis and plan his treatment. In the course of all of this, the owner has watched five or more white-coated individuals come in and consult at length about his animal.

Miss Romasco noted that although owners rarely initiated questions, some of them exhibited a considerable amount of discomfort, even anxiety, when they were faced with a whole cadre of young veterinary students. The summoning of a senior veterinarian escalated the anxiety for those who did not understand the procedure. Usually the senior veterinarian would explain the process when he arrived, but occasionally he forgot. Because of this, Miss Romasco arranged to stand near the owner when she entered the room, and while the student or students examined the animal she would explain who they were and what they were doing, introducing herself as well as the veterinary students. She also explained that the final diagnosis treatment plan would be done by the senior veterinarian. This simple procedure seemed to be much appreciated by both the veterinarian and the owners, for it tended to reduce the anxiety level and keep communication open. Miss Romasco would stay through the examination and be available for consultation with the owner if that seemed desirable. Occasionally, when the prognosis was poor or the animal had to be sent to X ray or to the laboratory, Miss Romasco would invite the owner or owners into the companion animal conference room to talk about their concerns and the possible decisions that lay ahead. As she gained confidence in her capacity to be helpful, she would go into the reception room and, if she observed signs of stress in any of the people waiting there, would introduce herself and ask about the pet. If in her judgement they needed her help, she would offer it. (Miss Romasco had considerable social work knowledge and experience on which to draw in order to make this kind of social diagnosis.)

In the early months of the clinic, case finding was a problem and most referrals were either from the oncology clinic or if they were from other parts of the hospital they dealt only with seriously disturbed individuals who were nuisances to the veterinarians.

SOCIAL WORK EXPERIENCE IN YEAR ONE—A SUMMARY

The social work service to the Companion Animal Clinic was offered for slightly more than six months of the first academic year—from late October through May. During that time the social work student was involved in twenty-nine cases for which more than a single brief clinic interview was required. She handled fifteen of these personally; fourteen were carried out in conjunction with one or more other staff members.

In most of these situations there were multiple problems to be dealt with. The categories we had started with were too broad to be useful for analytic purposes so, using a case problem analysis list from a human hospital as a guide, Miss Romasco developed the following list for the Companion Animal Clinic. It should be noted that although this carries the clinic name, it was primarily a social work list, and only two of the categories have to do with problems that are primarily those of animal behavior. (Veterinarians have observed that if they had been preparing it, the list would have had nineteen categories of animal problems and two of people problems!)

COMPANION ANIMAL CLINIC
CASE PROBLEM LIST

1. Clinicians suspect pet abuse.

2. Clinicians encountering owner created problems, e.g. constant telephone calls, threats, "acting out."

3. Clinicians and owners are having problems communicating, e.g. language barriers, emotional upset.

4. Owners have complaints about specific clinicians.

5. Owners have complaints about hospital procedures.

6. Owners need information and education about pet's illness and treatment and "translation" of medical terms.

7. Owners need information and education about hospital procedures.

8. Owners need help with informing other family members of the situation with the pet.

9. Owner's emotional reaction to illness, diagnosis, treatment, or prognosis of pet.

10. Owners need support and help in coping with pet's changed needs as a result of illness, e.g. pilling, special diets.

11. Owner needs help with immediate emotional reactions to death of pet.

12. Owner needs support and help in dealing with bereavement and grief reaction over time after the loss of a pet.

13. Owner's chronic anxiety, stress, or depressive reactions related to a pet.

14. Owner needs help in decision making around treatment of a pet.

15. Owner needs help in decision making around euthanasia of a pet.

16. Owner seeking information and advice on type of pet to obtain.

17. Owner needs information and advice on training and training school for a pet.

18. Undesirable behavior changes in a pet, e.g. aggressiveness, destruction, inappropriate urination and defecation.

19. Owner needs information and advice on animal behavior that is not necessarily a problem, e.g. introducing a new pet.

20. Owner has personal problems not related to the pet, e.g. family problems, interpersonal relationships.

21. Owner has concrete needs not related to the pet, e.g. finances, housing.

In addition to the subdivisions of earlier categories, some new ones have been added. The first one, identification of pet abuse, was one we asked clinicians to alert us to, not only because of the immediate problem but also because it is known that people abusers tend to victimize other members of the family if the first abuse object is removed. Sometimes, particularly in situations of child abuse, more than one person will be abused simultaneously. Operating on the principle that in many families pets fill a role of surrogate people, we felt that a responsible social service should have an opportunity to follow up on cases of suspected abuse. Only one such case was reported during this period, so no claim can be made to support the hypothesis; nonetheless, we believe this is an important matter for further cooperative study between social workers and veterinarians.

Many of the categories have to do with communication problems. I was involved in one such case in which there were obviously several parts to the problem. The veterinarian was using medical terms that the owner did not understand, and the owner was under so much stress from many problems of her own that she only half heard what she might otherwise have understood. The patient was a pretty little kitten with a degenerative neurological disorder of genetic origin for which there was no cure. The owner was an elderly woman who obviously cared a great deal for this kitten and for its two ill litter mates that she had been unable to carry to the clinic. The veterinarian, confronted by the woman's anxiety and his own sense of helplessness repeated several times a diagnosis and prognosis that I could barely understand because of the scientific terminology, though I did get the import of the message after listening several times to his careful explanation followed by the owner's response, "But he is getting better, isn't he?" I intervened, trying to capture the major points in lay terms. "You are saying that the kitten was born with a nerve disorder that cannot be cured but that, with the owner's care, it can be expected to live a couple of years longer if she is willing to hand feed it, put it in the litter box, etc." The veterinarian looked at me in some relief as he agreed my interpretation was correct, and the owner began to respond with some understanding of the situation.

I had approached this woman because she was crying as she sat waiting for the veterinarian to return from the laboratory where he had taken her kitten. As we talked she told me about all three kittens and their mother who had died earlier in the year. Then she began pouring out a series of other problems. She was terminally ill. Her husband had recently died. Her son had lost his job and his wife and children had left him. There was a lien against her house that she felt was unjustified, and she had no money either to hire a lawyer or to pay the lien. The kittens provided the one bright spot of her life, and she felt she could not bear the news that they might die. The news that they could have a couple of years was a great relief to her, though she held on to her hope for a cure.

Following this first session, Miss Romasco stayed with the case for several months, calling the owner periodically and helping her to find some community resources to assist with some of the other problems. Gradually this woman accepted the diagnosis and worked out a realistic plan for the kittens and herself. After several months she reported that she was able to function without clinic help but asked for permission to call back if necessary.

The Clinic has developed a procedure for following up clients at three-month intervals over a period of a year after a case is closed, and in our first follow-up call, our kitten owner was cheerful and seemed to be coping well for one with her problems. We were not able to reach her the second time and suspected that her own illness may have hospitalized her by this time.

An analysis of this case shows that problems fell in six categories:

3. Clinicians and owners are having problems communicating.

6. Owners need information and education about pet's illness and treatment and "translation" of medical terms.

9. Owner's emotional reaction to illness, diagnosis, treatment, or prognosis of pet.

10. Owners need support and help in coping with pet's changed needs as a result of illness.

20. Owner has personal problems not related to the pet.

21. Owner has concrete needs not related to the pet.

Communication

This kind of multiproblem case has not been unusual for the social work service. Numbers of problems in one case have ranged from one to seven, with an average of 3.5. Of the six most frequently seen problems, three had to do with communication. They were, in order of frequency, "owner's need information and education about pet's illness and treatment and 'translation' of medical terms"; "owners need information and education about hospital procedures"; "clinicians and owners are having problems communicating." The other three leading categories were "owner's emotional reactions to illness, diagnosis, treatment or prognosis of pet";

"owners need help in decision making around treatment of a pet"; "owners need help in decision making around euthanasia of a pet." In all of the cases there were six problems of owner need not related to the pet. Three were personal problems, and three were for concrete services such as finances or housing.

The number of cases encountered in these first months was small but the number and range of problems surprisingly large. They represent the kinds of problems a social worker might expect to find in any clinic setting where people are concerned about an ill family member. The one notable exception is that in human medicine there is no need to make a decision concerning euthanasia.

CONCLUSIONS, IMPLICATIONS, QUESTIONS

Letters and telephone calls from clients and word of commendation from veterinary clinicians have confirmed our impression that the social work service is a useful one. There are some owners who decline the service when it is offered, and there are still a few clinicians who are reluctant to refer cases or are not quite sure which ones to refer, but for the most part the service has been well used and valued. In the process of offering the service, staff has learned a great deal about the triangular relationship between veterinarian, patient, and owner and about the kinds of people that may appear in a veterinary office.

In spite of the frequently voiced opinion that a social worker would interfere with the relationship between veterinarian and owner, we have not had reports of this happening. Perhaps this is because referrals have been voluntary on the part of the clinician. Perhaps it is because social workers have not been aggressive in working with potential clients. Perhaps it is because the service is one that really enhances the relationship in a teaching institution.

The question of whether or not a social worker could be helpful in a private clinic remains unanswered. I suspect it would not be economically feasible to hire a social worker for a one or two person clinic, just as it is not feasible in human medicine. I would like to see the idea tested in a group practice arrangement, however, with a social worker handling intake, scheduling appointments, and billing clients, as well as being available for consultation and/or direct service in situations where people problems emerge. In smaller clinics, ways should be found to develop consultative relationships with human services personnel.

The first year's experience verifies our assumption that some people use a pet as a "ticket of admission," to create an opportunity to discuss other problems with a professional person whom they view as nonthreatening.

The premise upon which the Companion Animal Clinic is based has been amply demonstrated—that people and their pets form a system in which the mental and physical health of all members is intimately related. To deal with them adequately requires knowledge and skill from a variety of disciplines. In order to insure the best possible treatment, veterinarians should have access to psychiatric and/or social work consultation. Conversely, we believe that social workers and psychiatrists need more knowledge about the significance of pets and their behavior on the health and well-being of individuals and families.

CHAPTER 12

THE FIRST YEAR OF AN ANIMAL BEHAVIOR CLINIC
IN A VETERINARY SCHOOL: PROCESS AND OUTCOME

CEILE WASHINGTON

T he Companion Animal Clinic of the University of Pennsylvania Small Animal Hospital was established as an integral
part of the Veterinary Hospital to study the interactions between
pets and their owners seen at the hospital.

The clinic was to serve as both a training and research center
for veterinary students. Through the establishment of the clinic
we hoped to begin to answer questions about the social and psychological interactions between owner and pet and to define the
role of the veterinarian as a consultant in a multidisciplinary
team effort to diagnose and intervene in human mental health
problems. We also planned to collect data routinely on basic social
information, history of pet ownership, and the social relationships of pets in the context of the client's social relationships. The
clinic's major activities would include the diagnosis of a typical
interaction between pets and their owners and the diagnosis of
disturbances in animal behavior.

We initially thought that a large proportion of owners with pets
with disturbances in behavior would be people with mental disturbances, their emotional problems being manifested in their pets.
Also we hoped that this clinic would be useful in identifying
people who were in need of professional (psychiatric or social)
attention and who were using their pets as an indirect means of
seeking help for themselves. This assumption proved to be only
partially correct.

It soon became apparent that the reason why most people
brought their pets into the Companion Animal Clinic was because
the owners found the behavior of the pet disturbing.

Dr. M. J. McCulloch has suggested that the owner can play a significant role in inducing pet problems, many of which are manifested in the behavior of the animal. Pets allowed to roam freely have an increased chance of displaying intermale aggression. Some pets can develop attention-getting mechanisms such as sympathy lameness, coughing, or pruritus or can reflect the emotional state of the owner by exhibiting behavior such as hyperexcitability, housesoiling, aggression, and cowering.

In our society, adults are held responsible for the behavior of younger family members. We found most of our clients viewed their pets as younger family members. They often felt guilty in accordance with this belief that they had contributed to the development of distressing behavior.

We also found that the anxiety of the owner of a pet with a behavior problem was as great as that of owners of pets with organic illnesses. However, the nature of their anxiety was different. Owners were also stressed by the responsibility of deciding the fate of the pet since most owners were unaware that many behavior problems could be treated successfully. They believed humane destruction was the only choice.

When assured that an exploration into the chain of events leading up to the development of the disturbing behavior need not end in giving the animal away or in euthanasia, almost all of the clients seen by the clinic were willing to spend considerable time during the initial interview and in subsequent follow-up calls and return visits describing in detail their interactions with the pets. We requested that all "family members" who were regularly in contact with the pet attend the sessions so that we could collect the data we were interested in and could observe how each family member interacted with the animal and vice versa.

Each case was attended by the staff veterinarian and social worker. A review of earlier cases revealed that the data on basic social information, history of cat or dog ownership, and social relationships between the owner and the pet could be categorized. So two questionnaires were developed, one to be completed by the veterinarian (Fig. 12-1) and the other to be completed by the social worker (Fig. 12-2).

FIGURE 12-1.

BEHAVIOR FACT SHEET

Problem:

_____ Housesoils	_____ Barks	_____ No Obey
_____ Chews	_____ Aggressive	_____ Howls
_____ Digs	_____ Bites	_____ Shy
_____ Jumps up	_____ Runs Away	_____ Pica
_____ Unruly	_____ Fights	_____ Coprophagia

Other:

Problem notes:

Correction to date:

Dogs Reaction:

Age Dog Obtained: from:

Litter Behavior: Parents:

Houstraining Method:

Obedience Training:

Where sleep?

% Indoors (where):

%Outdoors:

Where spends the day   % Indoors (where):   % Outdoors:

Play Periods: Exercise:

Last Vet Check: Purpose: Outcome:

Other:

Diet: Daily feeds: 1 2 3   Who feeds:

Quantity: Supplements:

Other Pets (what kinds, what dates of acquisition)?

FIGURE 12-2.
FAMILY DATA

Persons Living In The House:

1. Adults: Number _____ Ages _____ Relationships:

2. ·Children: Number _____ Ages & Sex _____

Role of Animal in the Family:

Stressful Events or Changes in Family and/or Living Situations:

Problematic Relationship of Animal to Individual Family Members:

Differences Among Family Members in Handling Animal:

Impressions:

Although the veterinary clinicians at the Small Animal Hospital were willing to refer animal behavior cases to the Companion Animal Clinic, the actual number of cases that materialized were few. We had to find a means of increasing the number of referrals from other sources. The precipitating factor that increased the number of cases was unsolicited media coverage.

Early in May 1979, an advertisement was placed in a local newspaper seeking another social worker for the Companion Animal Clinic. A reporter took interest in this, interviewed the clinic staff, and wrote a story that appeared in the Sunday magazine section of a prominent local paper. Later in May, a second reporter from another paper featured the clinic in a human interest column. Television coverage by two stations followed. In·June, we sent letters to practicing veterinarians in a three state area announcing the availability of the Companion Animal Clinic for referrals.

The number of appointments made for the Companion Animal Clinic increased from about three or four per week to approximately ten to twelve per week.

SUMMARY OF CASES SEEN FROM
JANUARY THROUGH AUGUST, 1979

A total of 117 animal behavior cases were seen by the Companion Animal Clinic in the period from January through August, 1979. Of these eighty-five were canine behavior problems and thirty-two were feline behavior problems. Some owners brought in more than one animal at a time.

Among the canine population, twenty-five breeds were represented, including the category "mixed breed." The mixed breed pet was presented in twenty-three instances, far more often than any specific breed. There were nine terriers, five setters, and five retrievers following as breeds presented most often.

Males were presented more often than females. Intact males showed a higher rate of having behavior problems than castrated males. Of fifty-eight males presented to the clinic, forty-five were intact. Twenty-seven females were seen, with about half of them intact.

In classifying the canine behavior problems, the chief complaints were problems involving aggression, house soiling, destruction, and fearful behavior. The category of aggression was further broken down into aggression toward other dogs (forty-seven cases), aggression toward people (fifteen cases), and aggression directed toward both people and other dogs (three cases).

There were twenty-three dogs who exhibited some type of fearful behavior, such as a fear of thunderstorms or sudden, loud noises. In many cases, this was exacerbated by the pairing of a reward (increased attention from the owner) with the stimulus (loud noise).

There were twenty-two dogs who were presented for being destructive. This occurred most often when the owner left the dog alone for periods of time ranging from a few minutes to a few hours.

House soiling and nonresponsiveness to command were the next most frequent categories, followed by barking or howling, urine marking, mounting, roaming, and hyperactivity, in that order.

Of the feline cases, twenty were domestic shorthair cats, five

Siamese, three Persian, three Burmese, and one was an Abyssinian cat.

Equal numbers of males and females were seen. The number of neutered and nonneutered animals was equal.

Cats were presented most often for problems involving elimination or aggression. Urine spraying, inappropriate urination and defecation, and aggression directed toward conspecifics were presented in almost equal numbers. There was only one case of aggression involving a person.

We were able to identify some kind of environmental stress as a common denominator in most of the feline behavior problems. The situations that cats seem to find most upsetting are the addition of new family members, overcrowding with pets (cats or dogs), and changes in daily routines or environment.

The number of pets owned in a household containing dogs, does not seem to be important in evaluating who will be plagued by disturbances in pet behavior. Single pet owners have behavior problems in their pet almost as often as multiple pet owners.

One characteristic that was common among most owners of pets with behavioral problems was that no matter how disturbing the animal's behavior was, in most cases the owners felt that they still fulfilled a role in providing affection and constant companionship.

We routinely asked owners to describe what role their pet played in their lives. The most common response was that the pet served as a companion, followed by serving as a family member, most often a child. A few people reported owning the pet only for protection. One was a guide dog and companion. Only one person described the pet as a nuisance.

DIAGNOSIS AND TREATMENT OF BEHAVIOR PROBLEMS

Since many of the behavioral problems owners found upsetting are either normal canine or feline behaviors or learned behaviors it is necessary that the treating veterinarian understands how genetics, early experience, and learning contribute to the development and the display of these behaviors. Also important in understanding and treating disturbances in behavior is a working knowledge of ethology, psychology, and physiology because therapeutic approaches include changes in environment, hormonal and psychoactive drug usage, neurosurgery, behavior modification, and the treatment of organic disease processes.

Case Report One

The owner of a five-month-old female St. Bernard came in distressed and embarrassed by the fact that his dog would roll over and urinate whenever in his presence. This dog was acquired following the death of the owner's previous St. Bernard.

Because his previous dog had slept upstairs he tried to train the pup to do so by placing her forepaws on the first step, pushing from behind, commanding that she obey, and scolding her when she didn't. She would respond by rolling over and urinating. The owner would respond by scolding and beating. And the St. Bernard would respond by running away.

The owner's previous dog had been killed in a road traffic accident, and he wanted to train this pup to be road-sensitive and not roam. He would take her outside, show her the area where she was allowed to roam, and then remove her lead. She would immediately run away and on being caught would roll over, urinate, and receive another beating.

By the time we saw this pup the cycle of urination-punishment-escape-more punishment was so enforced that the pup would roll over and urinate at the sight of the owner.

Treatment of this problem began with a discussion about the differences in personality of individual dogs, allowing the owner to understand why the training techniques that were successful with his first pet were not necessarily the best techniques for his present pet. A discussion about the meaning of submissive urination in dogs followed. The owner agreed that he should discontinue his present method of training and begin to raise and strengthen his pet's confidence levels. He was given behavior modification techniques to use which were designed to build and strengthen the dog's confidence levels. Within two weeks he returned to demonstrate that he was able to scold, command, and discipline her without her showing any signs of submission.

Sometimes a combination of behavior modification therapy and psychoactive drug therapy are more effective.

Case Report Two

A very pleasant family of two adults and three children were extremely disturbed because the family pet, a two-and-one-half-year-old female spayed bassett hound began to be very fearful and anxious in the father's presence. She would shiver and run and hide if he tried to interact with her in any way. She would not eat in his presence but would accept food from him under the table. She would shiver and run and hide if he tried to pet her, talk to her, or approach her, unless he approached her backwards. She was friendly and playful with all other family members. This was upsetting to the family because they enjoyed doing many things together, including caring for the dog. The father was forced to remain a nonparticipant in this activity even though he reported that he loved the dog and would awaken before other family members to go down and feed and talk to the pet.

Although we could not identify the factors that precipitated this behavior, it was clear that it was fearful behavior. The dog was placed on antianxiety medication and the family was taught how to systematically countercondition her to exhibit a new response to the presence of the father.

The interaction between the pet and the father has improved significantly, and both are now happier in each other's presence.

DISTURBANCES IN OWNER BEHAVIOR
NOT RELATED TO THE PET

Occasionally we were presented with a case in which we suspected a disturbance in owner behavior not related to the pet. I say occasionally, but that is not to imply that few such cases exist. It may well lie in our lack of training to identify these problems. The following are examples of what we identified as such disturbances.

Case Report Three

A young lady in her early twenties was calling the Trauma Emergency Service of the Small Animal Hospital two to three times daily. She expressed great concern about the health of her cat. She insisted on knowing what would happen to her cat if it accidently ingested broken glass. No answer from students or clinicians would assuage her fears and she could not be persuaded to bring the cat in for a physical examination.

In the beginning, the students on duty in the Trauma Emergency Service tried to advise her of an appropriate course of action in the event that the cat did ingest glass, but they soon grew tired of her calls and began to tell her that the cat was going to die. Her fears increased. She persisted in calling and was referred to the Companion Animal Clinic.

During my first telephone conversation with the young lady I expressed my concern for her health due to her excessive worrying about the cat. She began to cry and admitted that she knew that her cat was in fine health and that she was very ill and could not stop herself from imagining various accidents befalling her cat if she accidently broke some glass.

Over the next several days the number of daily telephone calls to the Companion Animal Clinic increased to five or six. I learned that she was a patient at a community mental health consortium and was under the care of a psychiatrist. When asked if her doctor knew of her frequent calls to the clinic concerning her cat, she replied that she was not allowed to discuss the pet during therapy sessions. She asked that I help her because she did not feel that she was improving.

Her psychiatrist was called and was quite surprised to learn about the nature and frequency of her calls to the clinic and pleased to know of our interest. We discussed the situation with him and decided to discourage her calls for fear they would dissipate the effects of therapy with him but to contact him if they continued.

Case Report Four

We encountered a similar problem in the case of a young lady who came to the clinic accompanied by a four-month-old beagle. Almost all the visible parts of the owner's body were covered with scratches and bite marks. Her initial complaint was that she was afraid that there was something wrong with her pup's nervous system, causing it to attack her while she slept. We suspected that the problem was not an animal behavior one because of her emphasis on the fact that most damage was done to her breasts and genital region. She made no mention of the obvious damage to her face, arms, and legs.

A neurological examination of the pup revealed no abnormalities. We prescribed mild tranquillizers for the pup and explained several simple procedures for the owner to follow in order to teach the pup not to bite her. We also suggested that she not allow the pup to sleep with her for a short while. To this advice she became quite upset and explained that because she lived alone it was important that the pup sleep with her.

We asked her to return to the clinic, and on her second visit it was learned that she was under psychiatric care and that her problem with her pup was the subject of discussion with her doctor.

To avoid possible abuse we asked her to return the tranquillizers she had been given for the pup but to continue with behavior modification until the pup was old enough to be obedience trained. She continued visiting both the Companion Animal Clinic and her doctor, and the pup's behavior improved to the point where our services were no longer needed.

Other cases were seen at the Companion Animal Clinic where there was some question in our minds about the owner's behavior. In these cases, we could not say specifically that there were owner behavior problems.

The wide range of cases presented to the Companion Animal Clinic in its first nine months suggests that such a clinic is a viable means of introducing veterinary students to some of the problems that can arise from people-pet interactions.

As practicing small animal veterinarians, they will encounter situations similar to those described here. We hope that because of training in a companion animal clinic, students will be better equipped to understand clients feelings, attitudes, and concerns for their pets; that students will have a greater appreciation of the human-companion animal bond.

In our new curriculum, fourth year veterinary students will be able to work in the Companion Animal Clinic on a rota. Second year students will continue to assist senior staff in taking histories, placing follow-up telephone calls, and surveying clients.

In this type of clinic, the veterinary student will learn not to be afraid to ask clients to share their thoughts if they look distressed and to listen when the owner begins to share information about him or herself that may (or may not) be important in the diagnosis and treatment of the pet's problem. The problem may be such that the veterinarian cannot offer a solution but sometimes just listening is enough. The veterinarian should be prepared to know when and how a problem can be referred to some other member of a health care team.

By treating animal behaviour problems, students will be made more aware of the social and emotional significance of the pet. Certainly most students are aware to a certain degree of the meaning of the pet, but the diagnosis and treatment of a behavior problem is special in that it requires full exploration of the interactions of the pet and owner to use the knowledge gained from that in helping to facilitate a healthy and rewarding relationship between the two.

CHAPTER 13

GUIDELINES FOR PLANNING FOR PETS IN URBAN AREAS

ALAN M. BECK

E ver since people began to live in villages, there has been a relationship between animals and society. Today, animals are so much a part of our lives they have a place in our homes, recreation, and work. Villages are now often cities, but animals, especially pets, are still very much a part of our lives.

Pet ownership, especially dog ownership per household, is much greater in rural areas than in small towns, suburbs, or city centers. In the United States, 67.2 percent of farm households own dogs, compared with 38.0 percent of the households in small towns, 41.1 percent in the suburbs, and 30.9 percent in the city centers (Purvis and Otto, 1976). However, because of the great concentration of people in small towns, suburbs, and cities, nearly 91 percent of the animals are in urbanized areas, and only 9.2 percent of the dog population reside on farms.

In 1975, the United States dog and cat population was estimated at forty-one and twenty-three million respectively or nearly 33 percent of all households owning dogs exclusively, 12 percent owning cats exclusively, and 10 percent owning both. In fact, dog-owning households average 1.42 dogs, and cat households average 1.58 (Wilbur, 1976). The dog-to-human ratio in the United States is about 1:5.9.

Dog populations in the United Kingdom are estimated to be at 5.83 million for a dog-to-human ratio of 1:9.4; France has 2.42 million dogs for a ratio of 1:6.3, and West Germany has 2.42 million dogs for a ratio of 1:25. Since the late 1960s there has been an increasing trend in all these countries (JACOPIS, 1975).

In addition to the sheer abundance of pets, there is other evidence that they are an important part of the cultures of United

231

States and Europe. There is a long-standing tradition for some hundreds of years that protects pets against abuse and neglect. There is a vast financial empire associated with the sale of pets, pet foods and accessories (Nowell, 1978), and animals, especially dogs. In the United States, ownership is significantly associated with financial income, house size, and other indices of social success of the owner (Purvis and Otto, 1975; Schneider, 1975).

Nevertheless, despite the pet population's abundance, economic importance, and legal and social place in our culture, there has been virtually no physical planning to accommodate pets in urban design. In fact, the only planning appears to be a trend to restrict or eliminate pets in cities. The majority of new housing developments in the United States are associated with pet-prohibiting regulations. Indeed, 12 percent of former dog owners surrendered their animals because they were no longer permitted to keep them (Wilbur, 1976). Try to imagine the consequences of not planning for the automobile: no roads, no parking garages, or no regulations.

It is possible, however, to accommodate pets in cities in ways that are humane and equitable for all through legislation, environmental management and design, population management, education, and research.

LEGISLATION

The planning for pets must include regulations and the commitment for enforcement of laws that protect people and animals. As already mentioned, most developed nations have laws that protect animals against cruelty, however, their enforcement is not usually consistent. Apart from the inherent right animals have to such protection, I believe that animal abuse has long been overlooked as an indicator, monitor, and even precursor to the antisocial behaviors people inflict on each other, including child abuse and neglect, spouse beating, rape, and homicide.

Planning for pets must also acknowledge that pets are aliens in human culture and their management is necessary to protect them and people. Legal guidelines should include restraint or direct supervision of the animal when on public property. Loose pet dogs account for the vast majority of dog bites (Beck, Loring, and Lockwood, 1975; Feldmann and Carding, 1973) and in the United Kingdom are involved in 6 percent of all road accidents (Carding, 1969). In the United States, motor vehicle accidents account for 52.9 percent of the dogs and 16.3 percent of the cats presented at the Trauma Emergency Service of the University of Pennsylvania's Small Animal Hospital (Kolata, Kraut, and Johnston, 1974). Loose dogs in general enjoy significantly shorter life spans (Beck, 1973).

Another legal option is the licensing and identification of all dogs and perhaps cats. Licensing is a source of income, demographic data, and identification. Individual pet identification, like tattooing or inert implants as for cattle, would facilitate the return of strayed animals and better enforcement of pet abandonment and neglect. Licensing differentials could be used to encourage the keeping of sterilized pets and smaller dogs. Licensing programs could also be used to encourage vaccinations for rabies and distemper, and deworming.

Legal guidelines should address bite reporting, rabies surveillance, and guidelines for human rabies prophylaxis. Such guidelines should reflect local circumstances. In New York City, for instance, bites must be reported. However, they are rarely treated as a potential rabies exposure (Marr and Beck, 1976).

No legal guideline has received so much international recognition as New York City's canine waste law, the so-called "scoop law." Basically, New Yorkers are required to retrieve and dispose of their dogs' feces. Dog waste contaminates soil, inhibits grass and tree growth, and encourages rats and flies (Beck, 1973, 1979).

In addition, there are numerous studies establishing that dogs are frequently parasitized by *Toxocara canis* (Dubin, Segall, and Martindale, 1975; Burrows and Lillis, 1968; Anvik, Hague, and Rahaman, 1974), and failure to clean up after dogs seeds the environment with *Toxocara* eggs. Soil samples from varying locations in the United States have been from 10 percent (Dubin et al., 1975) to 20 percent (Dada and Lindquist, 1979) contaminated with *Toxocara* eggs, and in the United Kingdom samples from public places were 24.4 percent positive (Borg and Woodruff, 1973).

It is now widely recognized that the ingestion of embryonated *Toxocara* eggs can cause human illness, toxocariasis or visceral larva migrans (Cypress and Glickman, 1976; Shantz and Glickman, 1978; Gundy, 1979). The disease appears to have two forms, an intestinal migration or ocular involvement. The symptoms of the generalized visceral larva migrans vary with site of migration and diagnosis is difficult. The ocular form, which is more easily diagnosed, has been reported in nineteen countries (Brown, 1974) and now an enzyme-linked immunosorbant assay (ELISA) test has been developed that can measure blood antibodies against *Toxocara*, thus permitting us to objectively test for visceral larva migrans in the general human population.

Using this technique, 4 out of 1,000 five year olds were found positive in New York City, and other investigators in Scotland found 2 percent of 200 blood donors, 22 percent of 144 patients with ocular lesions, and 4 percent of 28 patients with hay fever or asthma symptoms were positive for *Toxocara* (Girdwood et al., 1978). Using an older skin sensitivity test, 4.1 percent of 170 children were found positive in the United Kingdom (Borg and Woodruff, 1973).

The canine waste law can be seen to have a sound public health value. In addition, less feces on the streets may also mean less infection in pet dogs. It should be noted that freshly passed feces contains eggs not capable of infection. If, by chance, eggs are ingested, they will be passed out of dog or human before the larvae can escape into the body. They require three to six weeks in the environment to embryonate.

Despite the possibility of disease, the major impetus for New York's law was the aesthetic insult caused by dog waste on the streets and parks.

There are two other findings on the New York experience: the city is definitely cleaner, showing that a moderate commitment to enforcement and a social expectation for compliance can change a cultural pattern. And there is no evidence at all that New Yorkers surrendered significantly more animals to the shelters or adopted less from the shelters following the enactment of the canine waste law (Beck, 1979).

Other legal guidelines to be considered should address the numbers and kinds (species) of pets permitted based on humane, conservational, and health considerations.

ENVIRONMENT MANAGEMENT AND DESIGN

Although a "scooping tradition" will go a long way in alleviating the environmental impact of fecal waste, urban planning should include dog exercise areas. So-called "dog runs" could be so placed as to be utilized without being a nuisance to the neighborhood.

In 1976, before the canine waste law, we conducted a monitoring and questionnaire survey of a dog run located within a university complex (New York University) in the Greenwich Village area of Manhattan. As an indicator of use we marked with spray paint and counted every fecal deposit over a twenty-four hour period and found only twenty-six new deposits in the run although there were 175 new deposits on the immediately adjacent streets on all four sides. It appears that only 12.9 percent of the people who came to the street on which the run was located actually had their dogs use it.

Interviews with people who preferred their dogs to use areas outside the run elicited the following reasons. Users did not consistently clean up after their dogs, and new potential users were afraid their dogs would get sick. A similar health hazard was perceived from the frequent puddling of water. There was no shade for dogs or people. The most frequent reason for not being a user was fear of dog fights.

Many respondents suggested separate runs for smaller dogs. We found no sex difference between owners or their dogs among users and nonusers. Most users lived within two city blocks of the run; a single user walked one-half mile.

In another study, also before the citywide canine waste law, we monitored leashing and scooping in a small city park where leashing and scooping were required.

The park was traditionally a "dog run." After several weeks of monitoring we initiated a one-week period of intense enforcement of the leash law and scoop law. After the week we returned to the normal sporadic enforcement.

Two observations were apparent. First, compliance, even with sporadic enforcement was correlated with age, i.e. older people tended to have smaller dogs on leashes and tended to clean up after them more than younger people, who tended to let larger dogs run free. Second, there was virtually no residual effect of intense enforcement, that is, the level of "noncompliance" returned to the pre-enforcement "background" within one-half day after we discontinued our intense effort. I suspect the success of the city wide scoop law is, in part, cultural inertia; the enforcement effort was greater, longer, and the media reinforced it. Now, even with less intense enforcement, the social milieu expects compliance. My point is that dog runs can be designed but require maintenance.

Basic guidelines for inner city dog runs should include the following. The area should be completely enclosed by a fence at least four-feet high. There should be a double gate entry so as to permit entry without animals escaping.

The surface should be hard, nonporous, well maintained, and sloped to permit flushing and drainage. The area should be at least fifty feet away from areas used by children or from adjacent occupied buildings. If occupied buildings are nearby, local wind currents should be considered.

Plastic or metal scoops and ample covered receptacles should be available to facilitate the collecting and disposing of fecal material. If the area is near occupied dwellings, it should not be used between 11:00 P.M. and 7:00 A.M. There should be an adequate maintenance program with consistent personnel and funding.

As an alternative to designated areas, urban planners may want to experiment with opening more remote portions of larger parks to be used by dog walkers or permit animals off leads for exercise before or after usual pedestrian activity, i.e. before 7 AM and after 10 PM.

With better than one-third of urban dwelling families owning dogs, it is time to plan the animal's need into urban design. New buildings could be designed with roof top or basement exercise areas, fitted with proper fencing, flooring, drainage, and trash receptacles. Old structures may also be suitable after careful evaluation of space and structural soundness of the roof. The point is there has been very little planning for pets other than some attempts to segregate them in housing or in parks.

A further environmental issue regarding animal waste is urban storm water runoff. Such runoff has been associated with the killing of street trees (Pivone, 1969) and the closing of shell-fish growing areas (Barnett, Esser, and Flatau, 1978). City designers considering a separate rather than a combined sewage system should be aware of the impact of urban animals (Geyer and Katz, 1965). Here again, scooping accompanied by appropriate disposal into the sanitary system would help alleviate the problem.

Another aspect of environmental planning is the area of land management. As a general rule loose pets are found in higher human density, lower socioeconomic urban areas, and are best managed by responsible ownership. Ownerless strays and feral dogs are more common in low human density urban areas, and such populations can be discouraged by the sealing of vacant buildings, clearing of lots, collection of abandoned automobiles, and clearing or fencing around garbage dumps or landfills.

POPULATION MANAGEMENT

Urban areas must plan animal control: facilities to capture, house, and euthanize unwanted pets and strays. As a general rule, about 20 percent of the owned dog population will pass through the sheltering facilities annually. Therefore, a city should estimate the anticipated yearly load and depending on length of minimum stay build a shelter of appropriate capacity. As a general rule, extensive holding periods are not cost-effective as a vast majority of captured pet animals are retrieved within the first forty-eight hours.

The field of animal control is rapidly becoming a necessity, and an appropriate commitment for appropriate facilities, funding, and trained staffing should be part of future pet-planning guidelines.

EDUCATION

Planning for pets should include educating the human population toward the reasons and regulations of animal management, responsible ownership, and encouraging the use of animal services such as humane shelters, veterinarians, and obedience trainers. Additionally, educational programs as part of the school curriculum and the media may address such topics as the choice of a pet, pet abuse, or even how to avoid being bitten (Beck, 1976). Many of the problems now faced by municipalities related to poorly supervised or unwanted pets are the result of a poorly informed public.

RESEARCH

The guidelines I have proposed are not much more than extensions of common sense; however, we know relatively little about the motivations of pet ownership (*see* Chapter 5), the importance of pets to the lonely or elderly (*see* Chapter 8), or just what is the "carrying capacity" of the pet populations, that is, the kinds and numbers of pets that can be ecologically and socially maintained without adverse consequences. How many veterinary practitioners should there be for routine and emergency services? Are zoonoses control programs adequate? Are there circumstances and populations of people for whom pets are not appropriate? There must be a greater commitment to research.

Planning for pets, including what might appear to be restrictive legislation, must not be viewed as antianimal. Quite the contrary: having no guidelines and no regulations invites problems. If problems become severe, then the ultimate guideline — no pets allowed — becomes and is becoming the acceptable solution. Acceptable guidelines make it possible for pet ownership to continue in a fashion that benefits both animals and people.

REFERENCES

Anvik, J. O., Hague, A. E., and Rahaman, A.: A method of estimating urban dog populations and its application to the assessment of canine fecal pollution and endoparasitism in Saskatchewan. *Can Vet J* 15(8): 219–33, 1974.

Barnett, C., Esser, A., and Flatau, A.: Animal waste: nonpoint source pollution. Suffolk County Soil and Water Conservation District. Nassau and Suffolk Counties, New York, 1978.

Beck, A. M.: The ecology of stray dogs. York Press, Baltimore, 98, 1973.

————: How to avoid being bitten by a dog. *Pet News*. 1(6): 18–20, 1976.

————: The impact of the canine clean-up law. *Environment* 21(8): 28–31, 1979.

Beck, A. M. Loring, H., and Lockwood, R.: The ecology of dog bite injury in St. Louis, Missouri. *Publ Hlth Rep*, 90: 262–69, 1975.

Borg, O. A. and Woodruff, A. W.: Prevalence of infective ova of *Toxocara* species in public places. *Brit Med Jr, 4*: 470–72, 1973.

Brown, D. H.: The geography of ocular *Toxocara canis*. *Ann of Ophthal,* 6(4): 343–44, April, 1974.

Burrows, R. B. and Lillis, W. G.: Helminths of dogs and cats as potential sources of human infection. *N.Y. State J Med,* 60(20): 3239–42, 1960.

Carding, A. H.: The significance and dynamics of stray dog populations with special reference to the U.K. and Japan. *J Small Anim Pract,* 10(7): 419–46, 1969.

Cypress, R. H. and Glickman L. T.: Visceral larva migrans: a significant zoonosis? *Mod Vet Pract,* 57(6): 462–64, 1976.

Dada, B. J. O. and Lindquist, W. D.: Studies on flotation techniques for the recovery of helminth eggs from soil and the prevalence of *Toxocara* spp in some Kansas public places. *J Am Vet Med Assoc,* 174(11): 1208–10, 1979.

Dubin, S., Segall, S., and Martindale, J.: Contamination of soil in two city parks with canine nematode ova including *Toxocara canis*: a preliminary study. *AJPH,* 65(11): 1242–45, Nov. 1975.

Feldmann, B. M. and Carding, A. H.: Free-roaming urban pets. *Hlth Serv Rep, 88*: 956–62, 1973.

Geyer, J. C. and Katz, L.: Combined sewers. In *Restoring the Quality of Our Environment: Report of the Environmental Pollution Panel, President's Science Advisory Comm.* U.S. Gov. Print. Off., Washington, D.C., 1965.

Girwood, R. W. A., Smith, H. V., Bruce, R. G., and Quinn, R.: Human *Toxocara* infection in west of Scotland. *Lancet,* June 17, 1978, p. 1318.

Gundy, P.: Rising number of man's best friends ups human toxocariasis incidence. *J Am Med Assoc,* 242(13): 1343–44, 1979.

Joint Advisory Committee on Pets in Society (JACOPIS). *JACOPIS.* Walter House, 418–22 Strand, London, 1975.

Kolata, R. J., Kraut, N. H., and Johnston, D. E.: Patterns of trauma in urban dogs and cats: a study of 1,000 cases. *J Am Vet Med Assoc,* 164(5): 499–502, 1974.

Marr, J. S. and Beck, A. M.: Rabies in New York City, with guidelines for prophylaxis. *Bull N.Y. Acad Med, 52*: 605–16, 1976

Nowell, I.: The dog crisis. McClelland & Stewart Ltd., Toronto, Canada or St. Martin's Press, New York, 1979.

Pivone, P. P.: What causes the demise of city street trees? *New York Times,* September 21, 1969, p. 41.

Purvis, M. J. and Otto, D. M.: Household demand for pet food and the ownership of cats and dogs: an analysis of a neglected component of U.S. food use. Staff Paper, Department of Agricultural and Applied Economics, University of Minnesota, St. Paul, 1976, p. 47.

Schantz, P. M. and Glickman, L. T.: Current concepts in parasitology: toxocaral visceral larva migrans. *New Eng Jr Med,* 298(8): 436–39, Feb. 23, 1973.

Schneider, R.: Observations on overpopulation of dogs and cats. *J Am Vet Med Assoc,* 167(4): 281–84, 1975.

Wilbur, R. H.: Pets, pet ownership and animal control: social and psychological attitudes. In *Proc Nat Conf on Dog and Cat Control.* Denver, Co, Feb. 3–5, 1976.

Purvis, M. J. and Otto, D. M.: Household demand for pet food and the ownership of cats and dogs: an analysis of a neglected component of U.S. food use. Staff Paper, Department of Agricultural and Applied Economics, University of Minnesota, St. Paul, 1976, p. 47.

Schantz, P. M. and Glickman, L. T.: Current concepts in parasitology: toxocaral visceral larva migrans. *New Eng Jr Med*, 298(8): 436–39, Feb. 23, 1973.

Schneider, R.: Observations on overpopulation of dogs and cats. *J Am Vet Med Assoc*, 167(4): 281–84, 1975.

Wilbur, R. H.: Pets, pet ownership and animal control: social and psychological attitudes. In *Proc Nat Conf on Dog and Cat Control*. Denver, Co, Feb. 3–5, 1976.

Zinkham, W. H.: Visceral larva migrans: a review and reassessment indicating two forms of clinical expression: visceral and ocular. *Am J Dis Child, 132*: 627–33, 1978.

CHAPTER 14

A CURRICULUM TO PROMOTE GREATER UNDERSTANDING

OF THE HUMAN–COMPANION ANIMAL BOND

LEO K. BUSTAD AND LINDA M. HINES

The people–companion animal bond can and should be an integral part of education from a child's earliest classroom experiences, through the university years, and in life-long learning. At all levels, the curriculum should develop in the learner particular qualities of mind and character rather than limited, time-dated skills. The most important of these qualities are —

- Sensitivity, which includes appreciation of and delight in animals.
- Understanding, which becomes more complex as the learner matures and is instructed in the various sciences.
- Involvement of the learner in the human–animal bond.
- Responsibility in human–animal relationships for the benefit of people and society.

In 1852, John Henry Newman in *The Idea of the University* (Newman, 1949) advocated teaching science or philosophy which imparted liberal knowledge with no practical, commercial, or professional end—knowledge for its own sake. He talked about "the principal, that all Knowledge is a whole and the separate Sciences parts of one." Newman stressed the importance of realizing the interrelatedness of all branches of knowledge (Fig. 14-1). "Hence it is that the Sciences, into which our knowledge may be said to be cast, have multiplied bearings one on another, and an internal sympathy, and admit, or rather demand, comparison and adjustment. They complete, correct, balance, each other."

In Newman's view, liberal education is the cultivation of the intellect to the end of intellectual excellence. It seeks to "open the mind, to correct it, to refine it, to enable it to know, and to digest,

241

Figure 14-1. John Henry Newman stressed the importance of realizing the interrelatedness of all branches of knowledge.

master, rule, and use its knowledge, to give it power over its own faculties, application, flexibility, method, critical exactness, sagacity, resource, address, eloquent expression." In such a university atmosphere, people from the different disciplines "learn to respect, to consult, to aid each other." The student, even though unable to pursue all branches of learning, nonetheless comes to appreciate the great outlines of knowledge and the principles on which it rests. This experience both informs the mind and builds character.

Too often in our educational programs, particularly in the sciences, we inculcate an expectancy of certainty of knowledge and an aversion and an intolerance for uncertainty. To foster the expectation of certainty of knowledge is a serious betrayal of the essence of the scientific method. In this regard, we have been using the wrong punctuation marks. It is wrong to symbolize science with a period—a full stop indicating or signifying finality. It is far more appropriate to symbolize science with a question mark signifying doubt, the need for a further look. The question mark best represents science as a powerful force for progress. Most of the scientific community has shied away from studying or teaching animal and human nature and behavior because it is "unscientific," meaning, of course, unsatisfactorily deterministic and marked with uncertainty. They counsel anyone with scientific aspirations to turn away from such questionable adventures and to devote attention to more gratifying and respected areas where certain knowledge is available in large supply.

We do not imply scientific facts as well as concepts are not important. Much basic information in our curriculum is like the multiplication tables—it must be learned, memorized, and used frequently to prevent disuse atrophy. But we need a greater appreciation for science in the broader perspective as well, stressing stimulation of critical thinking, problem solving, and intellectual curiosity.

The four previously mentioned qualities—sensitivity, understanding, involvement, and responsibility—will be emphasized throughout this chapter as we discuss all age categories in our curriculum, concluding with an example of what we have done—a possible model for others, adapted to particular situations.

CHILDHOOD EDUCATION

In the earliest years, a child's sensitivity to the world of animals is developed through reading and story telling. The child delights in fantasy world animals but through that fantasy world becomes aware of certain values, principles, and moral truths. Since the various animals are anthropomorphized, the students confront qualities that are both animal and human. Fortunate is the child whose early years have contained an introduction to Gene Zion's *Harry the Dirty Dog*; to the rabbit, cat, and walrus in Lewis Carroll's *Alice in Wonderland*; to Mole and his companions in Kenneth Grahame's *The Wind in the Willows*; to C. S. Lewis' lion and mythic characters of *The Chronicle of Narnia*; to E. B. White's remarkable Stuart Little the mouse, Charlotte the spider, and Wilbur the pig; and to Rudyard Kipling's Rikki-tikki-tavi in *The Jungle Book*. Children can also become aware of the impact of the animal world on the human being through hearing their native folklore. Listening to the tales of the Indian peoples of eastern Washington, Native American children grew up learning of Owl Woman, Coyote, and Salmon and how their actions, moral and amoral, influenced the humanlike community. As explained by *Humishumi* or Mourning Dove (Hines, 1976): "Vividly we recall old *S'whist-kane* (Lost Head), also known as Old Narciss, and how, in the course of a narrative, he would jump up and mimic his characters, speaking or singing in a strong or weak voice, just as the Animal Persons were supposed to have done. And he would dance around the fire in the tulemat-covered lodge until the pines rang with the gleeful shouts of the smallest listeners. We thought of this as all fun and play, hardly aware that the taletelling and impersonations were a part of our primitive education."

This process of making children sensitive to not only the animal world but the moral consequences of the actions of animals (and by implication, similar human actions) is an important first step in establishing the animal–human bond. A child's sensitivity to and appreciation for animals is further broadened through fictional accounts of animals in a more realistic setting. Lassie the dog, Flicka the horse, Miss Piggy of the Muppets, Snoopy the dog, and other memorable animal heroes and heroines interact with "real" children in novels and on television.

Concurrently, the curriculum for children can build upon the sensitivity, appreciation and delight being awakened in fiction and provide a sound *understanding* of the world of real animals. Many classrooms contain pets for the children's observation, often cages of gerbils or guinea pigs or tanks of fish (Figure 14-2). This is important because increasing numbers of children, especially in urban settings, are growing up without the animal contact that characterized a more rural society. According to the 1970 U.S.

Figure 14-2. Students in a Pullman, Washington, grade school gather around the classroom's pet guinea pig.

Census of Population and Housing, the percentage of increase in multifamily units (which usually ban animals) was over five times greater than that in single family units from 1960. Schools can fill the gap in understanding by teaching about the environment and characteristics of real, specific animals. Children read books on the life of a gerbil, the care of a dog, or the life cycle of the frog. As they begin to differentiate between the anthropomorphized animals of their fantasy and the real animals that surround them, their understanding of what these real animals do and mean to human existence begins.

Their understanding and natural curiosity can be broadened through actual *involvement*. Through sharing and experiencing contacts with real animals, the students reinforce the facts taught about the animals in their classroom. By routinely caring for animals in the classroom or the home, children learn the rudiments of *responsibility* of people for animals and also the bond from animal to human that comes from the quiet delight of holding a "warm furry."

It is important at this early stage that children have actual hands-on experience with several animal species (a variety of warm fuzzies). It is also important to expose them early to the importance of wild animals to society and the importance of letting them be wild and not trying to make them into pets. During these years, the attitudes of teachers and parents are critical in shaping the quality of the human–animal bond. The teacher's attitude toward nonhuman life is often imprinted on the children in the classroom.

This highlights the importance of incorporating course work that focuses on the psychological aspects of the companion animal–human bond in the teacher education curriculum. This is a critical dimension if we are ever going to properly inculcate into our young a healthy respect, care, concern, knowledge, and appreciation for animals.

Groups concerned with strengthening this bond can increase the quality of interactions in the schools between children and animals. In this regard, we worked with school administrators to conduct a survey in which teachers of children up to thirteen years of age in our local schools indicated the extent to which they deal with or would like to introduce material on animals or pets in the classroom. (*See* Appendix 1). Of the respondents, 30 percent indicated they teach animal/pet units or bring animals into the classroom. Over 60 percent indicated they would welcome assistance in this endeavor. We respond in three areas. We suggest ways in which pets can be introduced into the curriculum. We make resources available to them in the form of films and handouts selected according to their suitability for various ages and their sound subject matter. Or we send resource people into the classroom to talk with children about the animals that they are raising in their classroom, the pets that they bring in to visit, or the wild animal kingdom that they are studying. A trained resource person can greatly augment the teaching about the real world of animals and the value that they have for human life, especially if the classroom teacher feels inadequate about presenting such information or lacks the time to prepare such material. (We also conduct educational tours for all age groups in our College of Veterinary Medicine.)

HIGH SCHOOL EDUCATION

As children move into the teenage years, the curriculum necessarily becomes more complex but still should offer sensitivity, understanding, involvement and responsibility in the human–animal bond. Literature continues to play an important part in making the young adult *sensitive* to the human–animal bond. The Disneyesque fantasizing of earlier years is tempered by more complex knowledge of animals in relation to larger questions of good and evil and the youthful human consciousness. Through reading Ovid, Homer, Jack London, Herman Melville, or George Orwell, they can become aware of the mythologized relationship between people and animals, whether the animal be domestic or wild, real or symbolic.

Their *understanding* of the real world of animals expands in biology classes where they investigate the complexity of various species. At this level, students should receive an appreciation of science as a quest rather than just a cold set of facts to be memorized. Studies of the food chain and the ecological system make students further aware of the interdependence between men and animals.

In history courses, students learn about man's place in the history of our planet and gain perspective about this relative newcomer and the eons of animal life that preceded him. Special problems courses at the senior high school level can investigate the overwhelming tragedy of euthanasia in our society necessitated by the lack of human responsibility toward pets. Also addressed should be the question of what constitutes owner responsibility, from small things such as cleaning up after animals, to greater things such as restraint to prevent bodily injury or property damage by animals. Students at the young adult level are asking questions about what it means to be a responsible adult, and exploration of these questions should include consideration of people and animals as an integral part of the curriculum.

Opportunities for *responsible* animal care and *involvement* with animals need not cease simply because animals are no longer found in the classrooms as they were in the earlier grades. In the rural areas of the United States, children are intimately involved through the youth groups such as Future Farmers of America or the 4-H programs in raising animals, taking responsibility for their well-being, and coming to terms with the finiteness of both animal and human existence. (Often these same students care for animals used in supportive roles, for example, guide dogs for the blind.) (Fig. 14-3).

During the teenage years, teachers can build upon the idealism, sense of responsibility, and commitment to other people felt by many adolescents by involving the students in well-planned animal programs. By sharing their pets with people in nursing homes, by assisting in horseback riding for the handicapped programs as volunteers (Fig. 14-4), by confronting basic ethical questions about the ways in which people treat their animals, the students can make very real contributions to strengthening the human–animal bond.

UNDERGRADUATE EDUCATION

When we address the question of curriculum at the undergraduate university level, the arguments become very heated.

Figure 14-3. Guide dogs for blind persons are trained by 4-H youth such as Barry and Heidi Froseth in Pullman, Washington.

Usually there is no great resistance to introducing units on animals and human behavior in the preuniversity years. The biggest problem is locating suitable materials and resource people. However, in the university years, the battle lines form quickly as faculty debate what courses are important enough to occupy the student's time. President Woodrow Wilson, a one-time academi-

Figure 14-4. Horseback riding programs for the handicapped, such as this one in Bielefeld, Germany, depend upon help from youthful volunteers.

cian, observed that "reforming a college curriculum is as difficult as moving a graveyard." Hazard Adams in his book *The Academic Tribes* (1973) goes to the heart of the question when he says, "And then there is the stereotypical Principle of behaviour which says that ... *an educational principle is fine as long as it does not interfere with the departmental program.*" Whether in the humanities, the social sciences, or the biological sciences, every department is willing to say that flexibility in the curriculum is important as long as the other person is flexible. But of course, the classes required in their department are absolutely essential.

Adams pleads for the importance of humanistic learning, which he defines as "not a skill to be mastered *in order to* proceed to a higher skill. It is a continuous process, like life itself." He also talks about the aim of a liberal education:

> After all, as I have suggested, almost anything can belong to "general education," if we take the term literally. For that very reason, the term has always seemed to us unfortunate. By contrast, "liberal education" has always carried the emphasis of freeing and expanding the mind rather than offering a little information here and a little there. General education can live with the symbol of the academic pie for it is a horizontal idea; but liberal education requires a hierarchical image, and humanistic study finds its place at the base.

If the human–animal bond is to find a place in the undergraduate curriculum, support must be given to the idea of liberal education in its fullest sense. In that particular intellectual climate described earlier by Newman, it is possible to respect the importance of *understanding* both human and animal behavior and the interdependence and interrelationships between humans and nonhumans. The human–animal bond should be an integral part of the curriculum, not an addendum, for both scientists and humanists. Students acquire a *sensitivity* to this bond through philosophic, historical, and humanistic explorations of all life, human and nonhuman. Courses in psychology, anthropology, sociology, philosophy, history, and the humanities are the backdrop against which this bond can be explored by the best minds and supported by substantial and respected research.

In the university, as in junior and senior high schools, students can become directly *involved* in the human–companion animal bond through specific programs, and this should be encouraged.

At Washington State University therapeutic recreation interns, for example, participate in the Riding for the Handicapped program. Preveterinary and premedical students, psychology, humanities, and animal sciences majors are important members of the People–Pet Partnership Program, which will be discussed later.

PROFESSIONAL CURRICULUM

Faculty in our College of Veterinary Medicine confront and influence the undergraduate curriculum directly when they determine what constitutes legitimate requirements for admission to professional school. Both veterinary and human medicine usually find it necessary to require a preponderance of undergraduate courses in the "hard sciences." As Donald Warwick points out in his article, "Social Sciences and Ethics" (1977), the term "science" carries with it "connotations of objectivity, value-neutrality, experimentation, quantification, and prediction," and the scientific method is considered inherently superior to "the spongy speculations of philosophy." Any subject matter that fails to meet the rigid definition of "science" is, in too many minds, suspect as a legitimate part of a scientist's education at the undergraduate or graduate level.

Fortunately, this attitude is beginning to change, and the call is going out for more familiarity with and concentrated study in the humanities and social sciences. Robert Veatch (1977) reported in "Medicine, Biology and Ethics" that as of 1977, forty-two medical humanities departments existed in American medical schools and more were coming. In a 1972 study of trends in the human medical curriculum, Vernon Lippard and E. Purcell (1972) said reports from individual schools emphasized that the curriculum should encourage development of human qualities that will enhance the student's role as a physician, including "knowledge of self, ability to communicate, and willingness to assume responsibility." The curriculum should provide the student an opportunity to recognize the broad social and economic responsibilities of the medical profession. Various medical schools called for a broadened, diversified, and more flexible curriculum.

Much the same emphasis is evident in the very significant report by the National Academy of Sciences entitled *New Horizons for Veterinary Medicine* (1972). This report emphasized the need for flexibility in curriculum. One of the points made by the panel writing the report was that "the liberalizing elements of the curriculum should not terminate with the preprofessional program, in view of the evidence that an understanding of the humanities and social sciences is acquired more effectively as maturity and experience increase."

The underlying plea is that we reconsider what constitutes the well-educated member of society. We feel this well-educated person, especially if he or she is planning to become a scientist, is not best served by heavy requirements only in the "hard sciences." Rather, this person should receive thorough training in the social sciences and humanities as a necessary part of an undergraduate education and an important component of graduate or professional training. If the latter is the case, then we can expect to see in the entering classes of our health sciences schools more students who have the breadth of learning necessary for analyzing the causes and the effects of animal and human behavior. Their openness of mind will give them a greater vision of the value of companion animals in human life, and their "humane" education will help them address tough ethical questions in this realm.

Within the professional veterinary curriculum, some universities have found specific ways to explore the human–animal bond as part of a student's education. Dr. Michael Fox has articulated well the importance of the holistic dimension for the animal scientist who looks at the animal as a behaving entity, not just as a production unit. His article, "Animal Behaviour in Veterinary Medicine: Its Place and Future" (1970), makes a most convincing case for the importance of making courses in ethology an integral part of our curriculum. He believes the veterinarian should be taught "to think in a scientific way of the behaviour patterns of the animals with which he works." He stated that knowledge of normal behavior and the symptoms and etiological factors associated with abnormal behavior, in both food and companion animals, is essential for practicing veterinary medicine. In this regard Dr. Fox stated—

Unfortunately, many veterinarians are unaware of the basic principles of animal behaviour, let alone the best methods for rearing and training cats and dogs. Basic training in college could provide the small animal clinician with the knowledge to recognise not only the behavioural disturbances of pets, but also enable him to advise the owner on the correct method of upbringing and training pets and treatment of certain behavioural disorders with drugs (e.g., tranquilizers), deconditioning, and training procedures. By being aware of various relationships that can exist between owner and pet, he would be able to rapidly establish a psychological rapport with his client, and through advice and gentle handling alone, often avert severe disturbances (e.g. *anorexia nervosa*) as might follow separation of the pet from the owner during boarding in the kennels.

Part of the fault for this situation lies with faculty who defend the status quo as the only legitimate form of instruction. The price we pay in our educational processes is a failure in the development of clinical judgment and compassionate wisdom in our professional students, a slowness in gaining full appreciation and useful working knowledge in both the science and art of animal health care delivery, which has an important behavioral component.

Recently, Dr. Bonnie Beaver at Texas A & M told us that she knew of only about five colleges and schools of veterinary medicine in the United States that offer a course in behavior within the professional curriculum.* Fortunately, some veterinary and human medical schools have pioneered in offering courses that deal with the human–companion animal bond. Dr. Beaver described her animal behavior course at Texas A & M that focuses on small animals and covers fifteen weeks. The first week consists of general introduction, followed by three to four weeks on puppy behavior (socialization and expected behavior at specific ages). The next three or four weeks, students learn about adult behaviors, everything from communication to sexual, ingestive, and eliminative behaviors. The course then shifts to abnormal behavior for three to four weeks, covering the causes and treatment of such behavior and differentiating behaviors that are medically related or environmentally caused. During the last one or two weeks, the class looks

---

*Dr. Ben Hart at University of California, Davis, and Dr. Katherine Haupt at Cornell also offer courses in behavior.

at how the cat differs from the dog.

The University of Pennsylvania School of Veterinary Medicine offers courses entitled "People and Their Pets: A Social Perspective on Pet Owner Problems for the Veterinarian" and "The Dog in America: Social Victim and Sacred Cow." The former looks at stresses that develop when an animal becomes part of a human family system and which may surface in the veterinarian's practice as either health or behavior problems in the pet. The course presents a framework for understanding and working with the humans who affect and are affected by their pets. The latter course discusses the ecology of the domestic dog in America, evaluates its social roles and society's responses to them, and addresses the public health implications of a large carnivore cohabiting with people. In addition, Dr. Aaron Katcher directs a course entitled "The Nature of the Bond Between Humans and Animals," which is taught by six outstanding specialists.

Two veterinarians, Drs. R. K. Anderson and J. S. Quigley initiated a course at the University of Minnesota School of Public Health entitled "Animal–Human Relationships and Community Health." The goals are to stimulate students to examine and clarify their thinking on the nature of interrelationships of people and animals; to foster a more understanding and fulfilling personal relationship with animals; and, at the same time, to generate new ideas and approaches to deal humanely and realistically with serious people–animal problems facing society today (*see* Appendix 2).

At Washington State University, undergraduates can choose from four classes that touch on behavioral questions. "Introduction to Physiological Psychology" explores the functional relationship between the nervous system and behavior: integrated organ systems, sensory processes, and investigative procedures. "Advanced Physiological Psychology" presents neurophysiological, hormonal, and biochemical bases of regulatory behaviors. "Primate Behavior" focuses on laboratory and field investigations of behavior of nonhuman primates, emphasizing learning, memory, motivation, family structure, habitat, and behavior development. "Comparative Psychology" gives a comparative and operational analysis of some broadly represented animal behaviors.

A few of our veterinary students may have taken some of these courses, but they are not required or even recommended, as far as we are able to determine. We believe there are great deficiencies in our present curriculum. We would like to require preveterinary students to take introductory psychology that would discuss scientific methods in psychology, learning, memory, sensation, perception, and physiological psychology. They should also take a human behavior class which discusses personality development, social behavior, group interactions, abnormal behavior, and treatment. We would also strongly recommend a course in social psychology, which discusses attitude changes, conformity, interpersonal attraction, values, groups, and social influences. Hopefully, all of these classes would also explore relevant aspects of human interactions with animals.

At WSU, a course entitled "Health and Society" is presently taught to first year medical students. (We would like to modify this course for veterinary students.) The course seeks to increase the student's understanding of the components and dynamics of the health delivery system. The primary objective of the course as taught to medical students is to increase the student's knowledge of the following aspects of the health delivery system through assigned readings, videotapes, and discussion: (1) disease and the sick person, (2) the healing occupations, (3) the organization of health services, and (4) current medical and ethical issues concerning health care. (Part of the health care should be animal facilitated therapy.)

Another goal of this course is to increase the student's understanding of and sensitivity to the relationship between health and cultural/socioeconomic factors. Achievement of this objective is facilitated by requiring each student to conduct a health practices interview in the home of a low income, minority, or elderly individual. A veterinary student performing this assignment would certainly have a deeper appreciation for a client.

In a recent conversation with Dr. Michael Fox, we talked about the importance of integrating anatomy, physiology, and biochemistry courses to give the students a conceptual framework in which to place applied ethology. In their courses, veterinary students should consider the history of domestication of pets, the present uses of animals, and the reasons why people need pets.

We have made a beginning in our veterinary curriculum by teaching "Behavioral Mechanisms of Physiology," an examination of the physiological transduction mechanism that enables animals to interact behaviorally with their environment (*see* Appendix 3). Hopefully, more electives in the area of ethology can be made available to students during their second or third year of study in our Regional Veterinary Program curriculum. During their fourth year, we hope to introduce a client relationship class and a special problem option on selection, placement, and evaluation of companion animals (Bustad, 1980).

Both students and faculty must become involved in research programs in the behavioral sciences to augment such classes. We now have only limited research effort in behavioral areas. We would like to see an extension into the food animal, and equine areas, where we know so little about behavior as it applies to managing livestock and to proper handling and training of horses. We need to study the relationship of behavior and disease and parasites. In our understanding of animal behavior, we are not even rubbing two sticks together in our attempt to get a spark now and then. We are so primitive that we are only operating with one stick.

One of the most promising programs in recent years is the development of a Center for Interaction of Animals and Society at the University of Pennsylvania. Faculty and students from psychiatry, social work, veterinary and human medicine, behavioral sciences, and other areas collaborate for the benefit of pet and owner. Perhaps in time such teams will staff behavioral clinics at all colleges of veterinary medicine and in major cities.

EDUCATION OF THE GENERAL PUBLIC

The final area of curricular interest in the human–animal bond deals with the adult population no longer in our schools and universities (Fig. 14-5). The necessity for instilling in this population sensitivity, understanding, involvement, and responsibility is every bit as great as it is in children. However, the job is more awesome because we have no single way to reach people, most of whom were educated in schools with no curricular offerings in the area of human–companion animal relationships. One of the ways in which we have attempted to reach this population in our community of Pullman, to increase their understanding of the animals around them, is through our evening "Companion Animal Care Class." The class is conducted at WSU chiefly by our veterinary students, with some help from a few faculty members. In the opening session, we present the importance of animals to people. The course contains presentations on the care of a wide variety of popular species, as well as on the problems of choosing wildlife as pets. Poisonous plants are discussed, and one session is devoted to animal behavior.

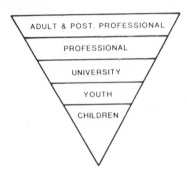

Figure 14-5.   Education that incorporates the human-companion animal bond is necessary at all five levels of our lives.

The general public's sensitivity to and responsibility toward animals can be awakened through public education programs in the media, in seminars, and in evening continuing education classes. The effectiveness of such presentations can be increased by using films and audiovisual aids, but most essential are teachers who are knowledgeable about and committed to the human–companion animal bond. Involvement of the adult population in programs involving companion animals can be very broad. They are essential as volunteers in horseback riding for the handicapped programs, in introducing pets in nursing homes and housing complexes for the elderly, and in bringing pets into various institutional settings (Fig. 14-6).

Training of lay volunteers is essential so that animal facilitated

Figure 14-6   Kathy Slinker introduces a kitten to a Moscow, Idaho, nursing home resident.

therapy programs are carried out in a well-planned and soundly conceived manner. Leaders of therapy groups utilizing the general public should be familiar with research being conducted in the universities and by practitioners in profiling animals, in constructing evaluative tools for potential owners and for institutional settings, in matching an animal to a particular owner or setting, and in evaluating the result of the introduction of animals into a particular situation. Use of such instruments is critical if the establishment of human–companion animal programs is to be taken out of the realm of the purely emotional and into the realm of the evaluative.

PEOPLE-PET PARTNERSHIP PROGRAM

Education at all levels is being affected in eastern Washington through the work of the People–Pet Partnership (PPP) Program. Guided by a council with members from many university departments and from most of the area's "helping agencies," the program has six goals:

1. Education at all levels about the responsibilities of pet ownership and the potential of pets for enriching people's lives.

2. Utilization of animals in therapy for the mentally disturbed and others.

3. Promotion of pets and companionship programs for the elderly, the lonely, students in dormitories, persons in prisons, and other institutions.

4. Establishment of a referral system to enable area residents to obtain specially trained pets (for example, handy pets for the handicapped or hearing dogs for the hearing disabled).

5. Establishment of a clearinghouse to provide information on pet programs and to link resource people with persons wishing assistance in the area of utilization of pets.

6. Formation of a consultant group that will work with those people who have specific people–pet problems.

The public's interest in this program is evident in the number of community people who come monthly for a two-hour education program. We started with ten people six months ago and now have over fifty. We have received many inquiries about our program from people locally, regionally, and internationally. The council's work in the junior high schools is summarized in the early section on Childhood and Senior High School Education. Undergraduate and graduate students at Washington State University are important members of the People–Pet Partnership Program and assist in introducing animals into classrooms, nursing homes, and elderly housing complexes (Fig. 14-7). The students, some of whom are handicapped, also participate in the preparation of evaluative tools to profile both the animals and the people they will help. They are part of the seminars sponsored by the People–Pet Partnership Program that bring together faculty in psychology, physical education, animal sciences, veterinary medicine, and other fields through the university, and they witness the fruitful interaction between people from various disciplines as they discuss problems that they all regard as important. These same faculty, along with community health science professionals, are guiding the establishment of sound programs in horseback riding for the handicapped, evaluation of animal facilitated therapy, and other areas. Eventually we expect to see courses, or parts of courses, in the university curriculum on animals in therapeutic recreation and in therapy, animal behavior, and assessment of animal–human interactions as a result of the work of the PPP. Several people have expressed an interest in pursuing postdoctoral research in animal facilitated therapy if funds are available.

Figure 14-7.   Nursing home resident in Moscow, Idaho, visits with a dog brought by a People-Pet Partnership council member.

Under the broad category of education for the "general public" we should also consider postgraduate learning. In order to make an impact on the immediate future, we should develop first-rate, autotutorial continuing education programs for professionals. These might adopt a variety of formats: slides and cassette tapes, programs for home computers, and videotapes. Such programs would enable a large number of professionals to learn about developments in the human–companion animal bond.

CONCLUSION

We have discussed in general terms both long– and short-term curricular changes that need to be made to address the human–companion animal bond. Although we have suggested some "add on" courses in the curriculum, significant lasting impact will come only from a reevaluation of the whole basic program of education. We designate from one-fourth to one-half of a lifetime as the necessary period for giving children and young adults the knowledge they need as adults in our complex world. During those formulative years, we must encourage teachers and administrators at all levels to give serious attention to the human–companion animal bond.

The potential for much harm as well as much good exists in human–companion animal programs, and it is essential that the balance is toward the good by providing sound leadership in these areas. The result will be a better understanding of the animals with which we are inextricably linked and an enrichment of the lives of many ages and conditions of people. Our primary program objective is to bring together disparate elements of the people–companion animal bond, elements that have been estranged too long, into a better and more lasting relationship. As the formal learning process from early childhood through to adult years incorporates material enabling sensitivity to, understanding of, and commitment and responsibility to the animals around us, we may witness the forging of new ties to each other as we learn to understand that enigma called "the human animal."

APPENDIX 1

People–Pet Partnership Program

What Is It?

The People–Pet Partnership Council is a group of people from area agencies who are promoting an awareness of the ways in which animals enrich our lives. The focus of the group is directed toward—

- Encouraging responsible pet ownership
- Utilizing animals in therapy for the mentally, emotionally, and physically handicapped
- Promoting companionship programs for lonely people of all ages
- Establishing a clearinghouse for information about pet and "helping animals" (hearing dogs, guide dogs) programs and facilities

How Can We Help You?

We need to know what is being taught in the public schools about animals. Also, we would like to offer audiovisual materials and assistance for your teaching in this area. Please answer the following questions and return this to the building secretary.

**Survey on Curricular Offerings
on Pets/Animals in Pullman/Moscow Schools**

(Grades K–8)

Teacher's Name _____

School _____

Grade Level _____

1. Have you taught a unit in the past on pets or animals?
   Yes _____ No _____
   If Yes, please indicate —

   Title of Unit _____

   Purpose of Unit _____

   _____

   _____

   Grade Level _____

   Resources Used —

   Books _____

   _____

   Audiovisuals _____

   _____

   People _____

   Animals _____

   Field trips _____

2. Do any of your other units have segments on pets or animals?
   Yes _____ No _____

   If yes, please explain _____

   _____

   _____

   _____

3. Do you have pets in your class room? (or have you had?)
   Yes _____ No _____

If yes, please state—

Species of pets _____

Number of pets _____

Length of time in room _____

Source of pets _____

Purpose of pets _____

_____

Problems with pets _____

_____

_____

Benefits of pets _____

_____

_____

4. Please indicate if you would like the following:

_____ Handouts on specific pets and their care

_____ Films on responsible pet care, pet problems, how to deal with aggressive dogs, etc.

_____ Field trips to pet/animal facilities, e.g. a stable, veterinary school, dog pounds, pet store, etc.

_____ Visits to the classroom by veterinary students, dog trainer, or other professional to talk about pets (with _____ or without _____ live animal presentations)

_____ Help with introducing a pet to the classrom and teaching about it's care or helping with problems

_____ Other_____

5. Would you like a PPP Council member to contact you about any of the items in question 4? Yes _____ No _____

6. Comments you would like to make about pet programs _____

_____

_____

APPENDIX 2

PERSPECTIVES: ANIMAL-HUMAN RELATIONSHIPS AND COMMUNITY HEALTH*

Class Topics

Cultural Perspectives
Between People and Animals—A Humanistic View
Owning a Pet—Ethological Considerations
Wildlife and People—Ethical and Ecological Aspects
Psychology of Having a Pet
Animals and People—An Animal Welfare View
Animals and People—A Public Health View
Animals and People—A Biomedical Research View
Animals and People, Issues and Problems—Some Ethical Considerations

*Compiled by the University of Minnesota School of Public Health.

APPENDIX 3

Behavioral Mechanisms in Physiology*

Introduction

1. Introduction to course and neuroanatomy
2. Neuroanatomy and behavior
3. Neurotransmitters and chemical manipulation
4. Neurochemical pathways in brain

Motivation and emotional behavior

Brain substrates of reward and motivation
5. Pharmacology and neurochemistry of reward
6. Anatomical studies of reward
7. Endorphins, narcotic receptors, and behavior
8. Stimulus bound behaviors and reward pathways

Emotion
9. Physiological indicators of emotion
10. Hormonal controls of affect
11. Neuroanatomical substrates of emotional states
12. Stress

Psychopharmacology
13. Behaviorally active drugs and the CNS I
14. Behaviorally active drugs and the CNS II

Substrates of behavioral abnormality
15. Neurochemical theories of schizophrenia
16. Biogenic amines and depression
17. Endocrine involvement in depression

Physiological control of ingestive behavior

Eating and energy
18. Evidence of control and the role of the GI tract
19. Glucostatic theory and the glycoprivic control
20. Body weight, fat, and feeding

---

*Compiled by the Washington State University College of Veterinary Medicine.

21. Neurotransmitters and control of kidney
22. Human feeding and obesity
23. Cancer and feeding

Thirst and body fluid homeostasis

24. Cellular thirst—osmoreceptors vs. Na^+ receptors
25. Extracellular thirst—ECF volume and the role of the kidney in behavior
26. The location and role of angiotensin receptors in thirst.

Sleep

27. Physiological and electrophysiological correlates of sleep states and stages.
28. Sleep centers and pathways
29. Is sleep necessary?
30. Sleep disorders
31. REM and arousal mechanisms

Hemispheric specialization and lateralization of function

32. Intuitive vs. analytic hemispheres
33. Lateralization in nonprimates (bird song)
34. Neural lateralization and specialization in nonhuman primates
35. Split brain studies in humans.

Sexual behavior

36. Differentiation of gender
37. Hormones and adult behavior
38. Sex pheromones

Neural plasticity and recovery of behavioral function

39. The classics
40. Neuronal correlates of recovery
41. Regeneration of olfactory neurons and recovery of olfactory guided behaviors

Aging: Some neural and behavioral correlates

This is an elective course.

REFERENCES

Adams, H.: *The Academic Tribes.* Liveright, New York, 1973, p. 45.

Bustad, L. K. *Animals, Aging and the Aged.* Univ of Minn Press, Minneapolis, forthcoming 1980.

Committee on Veterinary Medical Research and Education, National Research Council: *New Horizons for Veterinary Medicine.* National Academy of Sciences, Washington, D.C., 1972, pp. 80–81.

Fox, M. W.: Animal behavior in veterinary medicine: its place and future. *Vet Rec,* 86: 678–82, 1970.

Hines, D. M. (Ed.): *Tales of the Okanogans.* Ye Galleon Press, Fairfield, WA, 1976, p. 12.

Lippard, V. W. and Purcell, E. (Eds.): *The Changing Medical Curriculum,* 2nd ed. Josiah Macy, Jr. Foundation, New York, 1972, p. 11.

Newman, J. H.: The idea of a university. In Grebanier, B. D. et al. (Eds.): *The Victorian Age.* The Dryden Press, New York, 1949, pp. 689–98.

Veatch, R. M.: *Medicine, Biology and Ethics.* The Hastings Center Report, Special Supplement, 7(6):2, 1977.

Warwick, D. P.: *Social Science and Ethics.* The Hastings Center Report, Special Supplement, 7(6):9, 1977.

SECTION IV.

VETERINARY PRACTICE AND THE HUMAN–COMPANION ANIMAL BOND

CHAPTER 15

ATTACHMENT BETWEEN PEOPLE AND THEIR PETS:
BEHAVIOR PROBLEMS OF PETS THAT ARISE
FROM THE RELATIONSHIP BETWEEN PETS AND PEOPLE

VICTORIA L. VOITH

Having always been interested in animals, having been in a household with pets throughout most of my life, and having treated animal patients as a veterinary practitioner, I have always been aware that people are attached to pets. But it was not until I began working with people concerning the behavioral problems of their pets that I became dramatically aware of how strong the attachment of people is toward their pets.

Working with behavior problems of animals, a clinician spends several hours in discussion with the owner or hopefully the entire family. These discussions may span several weeks or occasionally a few months. During this course of time, the practitioner develops an impression regarding the persons' philosophies, rationality, personal and environmental interactions.

When I began working with animal behavioral problems, I had little idea what range of behavior problems I would see or what combination of techniques I would use. But I did have expectations regarding the people. I thought I would primarily see people who had not had pets before, who had not taken their dog to obedience school, and perhaps many people who were overtly unrealistic in their attitudes toward their pets and life in general. These are the "neurotic" or problem owners that veterinarians find difficult, tiresome, or aggrevating to deal with. It certainly would not be surprising to find animals with behavior problems in such households. But these were not the type of clients I saw.

The people I have seen have usually had pets before, pets of the same species or even the same breed. The former pets did not exhibit the same behaviors that the present pet was engaging in. Many of the dogs had been to obedience school and had done well. The people themselves appeared to be quite normal. No apparent neuroses, psychoses, or idiosyncrasies. If anything, my clients seemed to be better educated, more intelligent and rational than the average owner I saw across the table as a regular practicing veterinarian. The clients were, however, very attached to their pets.

Exactly what attachment is is currently being debated (Blurton Jones, 1972). Is attachment a motivational entity or a set of specific behaviors?

For the purpose of this chapter, attachment is considered an entity—an emotion or affective state that causes an individual to keep another in proximity or in frequent communication, and that results in physiological and behavioral responses by the former when the individuals are separated (Bowlby, 1969). People report that these responses are accompanied by unpleasant feelings. The discomfort and responses are alleviated when the individuals are again in proximity. And attachment, at least among human beings, can result in altruistic behaviors.

How do we know people are attached to pets? People take measures to keep pets close to them. Pets are often kept in the home, are taken on trips, accompany owners wherever they go throughout the day, and may sleep with the owners. An owner may curtail a visit or return home sooner than otherwise because a pet is at home. An owner may sever a relationship with another person because the friend does not like the pet. Owners also engage in considerable effort and financial expenditures to ensure that their pet has a happy and prolonged life. Most owners experience some grief when their pet dies, and there have been reports of profound grief when a pet dies (Keddie, 1977).

Owners may do some of these things because they miss the company of the pet. But sometimes owners also structure their activities around the animal because they believe that the pet will also miss them, and the owner wants to spare the pet any unhappiness.

Pets indicate that they are also attached to owners. An animal may follow an owner within a house, always staying in the same room as the owner. When separated from an owner, it may howl or engage in escape behaviors. Some pets become very quiet, stop eating, and look despondent when there is prolonged separation from an owner.

CASE HISTORIES

The following case histories provide examples of attachment people have for their pets. I emphasize that I am not reporting these cases because I believe the owners are pathologically disturbed in any way. I am reporting them because they exemplify the attachment owners can have for their pets.

Case One

Mr. M. presented his intact two-year-old male Weimaraner for help in curbing the dog's destruction of household items, in fact, of the house itself, when the owners were absent. The dog was very active in the examination room. He was constantly walking around, pestering people, or straining at the end of the lead. The owner, in a joking way, said at the beginning of the interview that this dog had cost him about $3,000. The dog had chewed up furniture, woodwork, and carpeting, and the owner had also incurred medical expenses to correct the consequences of the dog eating undigestable and/or dangerous items. The owner said he could live with the hyperactivity of the dog and the fact that the dog did not come when called but was rapidly approaching the point of not being able to afford the dog. The owner was also concerned about self-inflicted damage as a result of the dog's ingestive tendencies.

When asked why he had kept a dog such as this for as long as he had, Mr. M. answered, as he grinned with one arm circling the dog and the other petting the dog, "Cuz I love 'em." Mr. M. then launched into a discourse about the lifelong responsibility a person has when one acquires a dog. But his immediate response had been one of affection toward the dog. Whenever the dog was near Mr. M. during the interview, Mr. M. would stroke the animal. He was clearly attached to his dog and had borne considerable expense and discomfort rather than part with his pet.

Destructive behavior of dogs is a common behavioral complaint. Many owners put up with varying degrees of destructiveness for years because the pet has other redeeming values and the people are attached to it.

Case Two

The F. Family and their neighbor Dr. B. sought advice concerning the behavior of the F. family's dog, a one-and-one-half-year-old, ninety-pound, mixed breed, spayed female. They were concerned about the probability of the dog attacking people and/or killing other dogs. The F. family lived on a section of land bordered by Dr. B.'s land. Access to the F. family's house was through Dr. B.'s property. Almost everyone in the surrounding semirural and wooded community had two dogs that ran loose. One of the F. family's dogs, the ninety-pound female, killed a dog belonging to Dr. B. This incident took place on the doctor's property near the boundary of the F. family's land.

Although the history and temperament of the dog did not indicate that she was predisposed to injuring a person, it was quite conceivable that she would again attack other dogs. The F.'s decided to take precautions with this dog, but Dr. B. was still not convinced that the dog was not a threat to his children and those of the F. family. There was a constant insinuation by Dr. B. that the F.'s should get rid of their dog. Instead, the F.'s decided to move.

People will frequently move rather than part with a pet. Complaints from neighbors, intolerant roommates, or a change in leasing policies may force the issue. Even though it is inconvenient to move and the owners would rather stay, they move in order to keep a pet.

The previous cases involve monetary expenses and inconveniences in keeping a pet, but the following cases involve animal behavior problems that are a greater liability.

Case Three

Ms. X. presented her dog, a thirteen-year-old intact male poodle, for help with a bizarre biting problem that had been occurring for several years. The dog slept on the same bed with the woman, and while the woman was sleeping the dog would occasionally bite her on the back or arm. The bites usually resulted in bruises, never in puncture wounds or lacerations. Ms. X. could not correlate the biting with any consistent event except that she was always asleep when the bites occurred. The bites did not happen in any temporal pattern of which the owner was aware. The owner did not remember moving or touching the dog before he bit. The biting was what would wake her up. Her response was to ignore the dog and go back to sleep. She said the dog would not stay on the floor during the night and if confined or restrained would bark incessantly.

The dog had other objectionable behaviors. It occasionally resisted grooming and objected to being dried off by snapping or growling. The owner would get very angry at the dog in such situations, but she was not an effective disciplinarian. Primarily she was afraid of being bitten by the dog in such circumstances.

Ms. X. was obviously concerned about the problem or she would not have sought help. When asked why she had kept the dog, she answered that she did not believe in euthanasia which, to her, had seemed the only alternative. She did not think anyone else would take a dog that exhibited such behaviors. Besides, she said, the dog was probably not going to live much longer, and she was attached to it.

Ms. X. was in her 30s and lived alone except for the dog. She was very attractive and vivacious. Her personality was friendly and very warm. She obviously liked people, and almost anyone would immediately like her. She was a teacher and had recently won a teacher of the year award in a fairly large metropolitan school area (city of 300,000). She had authored well-received books regarding specific behavior problems of adolescents.

The nocturnal biting problems were alleviated by medicating the dog with low levels of barbiturate prior to bedtime. It could not be determined whether the effectiveness of the drug therapy was due to suppression of epileptic fits that occurred during sleep or merely induced a deeper sleep in the dog so that he was not disturbed during the night.

The owner chose to avoid or put up with the other behavior problems of the dog, e.g. resisting grooming, rather than invest the time necessary to correct them.

This particular case reflects a person's regard for a pet that exceeded regard for her own physical well-being. She had put up with the behavior because she had believed that the only alternative was to kill the pet. Such situations are not unusual. Owners frequently tolerated being bitten by a pet because they believed the only solution to the problem was euthanizing the animal. And they love, a term they frequently use, the pet, so they incur the injuries rather than exterminate the animal.

Case Four

Another interesting case involved a four-and-one-half-year-old uncastrated male bull terrier. A few months prior to being presented to the behavior clinic, the dog suddenly began staying upstairs in the bedroom areas rather than being on the main floor where he previously had spent most of his time with family members. Shortly after this he suddenly attacked, at independent times, the two men in the house. The family consisted of a married couple in their 40s and their seventeen-year-old son. The attacks were sudden, not accompanied by growling or piloerection. They were, however, paired with the occurrences of a loud, high-pitched noise, such as gunfire on T.V. or the sound of a chair scraping across the floor. The attacks occurred at approximately two-week intervals. The husband was so severely bitten on the hand that he was unable to work for two weeks.

The dog seemed very tired after the attacks and totally unemotional. He did not appear remorseful or persistently malicious.

The dog gave no indication of competing for a resource or asserting dominance over any of the family members. They had had this pet for four and one-half years without any signs of aggression. All family members could take food away from the dog, push him, pull him, pick him up, restrain him, trim his toenails (a manipulation he did not like), displace the dog from his resting place, stare at him, and order him out of the room, all without evoking aggression. He had frequently been disciplined, especially by the mother, by having his muzzle held and shook. In none of these circumstances had he ever growled, bared his teeth, or attempted to bite.

The dog had always been very much treated as a member of the family. He had been taken on a cross-country camping trip a year ago and previously had always been in the same room as the people. The dog often engaged in endearing "clowning" antics that amused the family. The wife, who had been periodically bedridden for several months and with whom the dog spent the day, said she felt toward the dog as though he were a son. The husband was also very attached to the dog. He took the dog on frequent walks. During the office visits he petted the dog frequently.

The dog's attitude toward strangers was unaggressive. The dog would bark at the door when a visitor approached and when the door was opened would greet people in a friendly manner. While in the veterinary hospital the dog had blood samples taken, neurological and ophthalmological examinations, and was repeatedly picked up to be placed on tables. He never growled or attempted to bite in any of these circumstances.

The owners were aware of the risk of keeping such a dog while we attempted to arrive at a diagnosis and treatment regime. They were also aware that the behavior may not be able to be diagnosed or treated with a sense of confidence. No guarantee could ever be made that the dog would not bite again.

They were willing to take the risk in the hope that a definite diagnosis could be found and the condition treated. We were working on the premise that the behavior was an audiogenic psychomotor seizure.

During subsequent weeks, as various diagnostic tests and drug therapies were tried, the dog severely bit the son and attacked both adults but was subdued without causing injury to the adults. The attacks were usually associated with a loud, distinct noise.

The owners were taking precautionary measures of usually muzzling the dog when he was with them or putting the dog behind a gate to keep him in another room.

The owners expressed some embarrassment about keeping the dog and trying all these measures. They admitted they were apprehensive about being around the dog. They rationalized keeping the dog rather than euthanizing him because he had been a good pet for the four and one-half previous years and was such a good, lovable dog between the attacks.

The only abnormality detected in the battery of tests performed on the dog was a moderately elevated blood urea nitrogen. Independently a low protein diet and primidone therapy were instigated. During this interval the dog had attacked the son and severely bit him on the arm. He also attacked the husband and wife but both were able to subdue the dog. These attacks were associated with noises or sudden sound apparently audible to the dog. The family was still interested in treating the dog.

Eventually the dog was placed on high doses of diazepam. After several weeks of remission, the dog, unmuzzled, twice suddenly ran into a room and threatened the mother who managed to fend off the dog without being bitten. These attacks were not associated with any noise audible to the family. Subsequent to the attacks the dog was tired and later friendly.

Following these attacks, the owners decided the episodes were becoming less predictable and were not being sufficiently controlled. Hence the risk to themselves was greater. They elected to have the dog euthanized. Both husband and wife had tears in their eyes when they brought the dog into the hospital. The dog spent the night in the wards and endeared himself to the ward personnel who could not believe the dog was to be euthanized because of aggression.

Case Five

The Z.'s had been successful in preventing their dog, a three-year-old spayed female Doberman, from manifesting any signs of aggression toward their young son for the past six months. Previously the dog had growled or snapped at the boy who was then approximately eighteen months old. Recently, in an unsupervised situation near the dog's food bowl and behind a door through which the mother had just passed to the basement, the dog bit the child on the face. The owners, accompanied by the child and dog, sought help concerning the dog's behavior.

Because of a combination of factors the owners were discouraged from keeping the dog. They were told that there was little likelihood of significantly reducing the probability of the dog biting the child in uncontrolled circumstances.

The child was a very active boy who periodically ran a small toy car over the dog's body during the interview. The dog looked apprehensive and somewhat frightened whenever the boy approached her. The dog gave no signs of aggression during the visit and occasionally solicited petting and contact from adults in the room. The dog had a "wary" look about her and appeared "nervous." Noises elicited quick and sudden orienting responses.

During the office visit the dog suddenly engaged in a very strange behavior, a behavior that the owners said the dog frequently exhibited but that they had not thought of mentioning.

The dog had grabbed the skin over her pectorals. She could be distracted from this posture by pushing on her, prying the mouth open, or making a loud noise. She demonstrated this behavior bilaterally. The owners said she would engage in this behavior for hours at home in either a standing or lying position and in a variety of circumstances.

The owners were distraught at the poor prognosis. The woman had tears in her eyes as she discussed what the alternatives were. She repeatedly talked about how loving, playful, and gentle the dog was. Both adults petted the dog frequently during the visit. They said the dog was a good watchdog, a trait they very much appreciated and a trait they thought would allow them to find the dog a new home without children.

Euthanasia was discussed, but the owners could not permit that. They left determined to find the dog another home. Because a clinician was interested in possible cessation of the pectoral sucking behavior in response to psychotropic drugs, the owners were given a prescription for two weeks of medication for the dog. If the owners placed the dog before all the medication was used, they were to ask the new owners to continue medication and if possible to put the hospital in contact with the new owners.

The owners were called a month later. They still had the dog. They had not been successful in finding the dog a home but were still looking. The risks involved in keeping the dog were acknowledged. The owners were taking precautions of always supervising the dog and child together. There had been no aggressive incidents since the last visit. The dog was still grasping the skin over the pectoral region. In fact, she had been engaging in the behavior prior to and during the telephone call.

Not infrequently, people will keep pets that are periodically injurious to their physical well-being or even to their children's health. These people do not appear to be masochists or sadists. They would prefer the animal was not injurious. They are embarrassed about the pet's behavior. Usually they have considered getting rid of the pet, but that generally means euthanasia and the owners cannot bear that alternative because they love the animal. They would give the pet away if they could, but in reality it is difficult to find a home for a behavior problem dog. The dilemma experienced by these people is quite stressful.

By far, most of the people I have seen struggle with this decision appear to be reasonable, healthy adults, who cognitively recognize the dangerousness of the situation and the logic of getting rid of the pet. But the attachment to the pet is so strong that they have difficulty, great difficulty, making the decision. When they do so, the decision is wrought with grief. Some, instead, assume the risks because they are unable to destroy the pet.

WHY THE ATTACHMENT?

One cannot repeatedly see such cases of attachment without speculating about why it exists. How does it develop and what maintains it?

Intraspecific attachment seems readily explainable. Attachment and behaviors surrounding attachment serve to keep individuals together and as such are mechanisms for social cohesion. Attachment behaviors would be inherent in social animals, such as people, many canids, and equids. The benefits of sociality are tied in with survival for these individuals.

Attachment to an individual other than one's own species is not as easily understood, unless the affiliation is a necessary symbiotic relationship. Although pets can provide benefits for many individuals, the human–companion animal bond is not universally essential to people. Many people live very complete, satisfying lives without pets. Then why are so many humans able to so easily form such compelling attachments, attachments that are not essential for survival and in fact are costly and perhaps even detrimental, especially when one's own offspring are threatened?

One way to find such answers is to look at mechanisms of attachment. Just as attachment behaviors are mechanisms to maintain social groups, there are mechanisms to maintain or enhance attachment.

The mechanisms of intraspecific attachment among human beings are only beginning to be investigated. Some of the parameters being considered as possible mechanisms of attachment are proximity, duration of proximity, care-soliciting behaviors, caregiving behaviors, dependency, responsibility for, sharing of experiences (especially happy ones), cooperative behaviors, feelings of joy or happiness evoked by behaviors of another, and tactile stimulation.

A joyful greeting evokes a warm response from the returning person and may enhance the attachment the returnee feels toward the greeter. A demonstration of sorrow or unhappiness at the departure of someone will usually evoke an emotion reaction in the departing person. This response may lead to the person returning sooner than otherwise. The sequence of behavioral responses and emotions may also strengthen the attachment.

An emotional demonstration, be it happiness at return or distress at departure, is a signal that the abandoned individual cares or is attached. These demonstrations tend to evoke similar feelings in the receiver of the signals. Demonstrations of being loved or being of value are quite likely to enhance attachments that already exist.

Children have additional mechanisms to evoke attachment. Sign stimuli of rounded features, specific reaching and smiling gestures are present to enhance attractiveness.

Since children are one's genetic representation in the future, and if attachment is a mechanism to enhance survival of the young, attachment is likely to be very strong between parent and child. And such mechanisms of attachment ought to be relatively easy to evoke. Since mothers have a considerable biological investment in their children and mothers are usually certain that a child is actually hers, the attachment of a woman toward her child should be very strong and relatively easy to establish and maintain.

All these traits and circumstances fit well into a human social system and are possible mechanisms for keeping a family unit together or, shall we say, attached. Pets fit in equally well, if not better.

Cats and dogs are usually raised from helpless infanthood by an owner and are generally quite dependent upon the owner for survival, not only as a youngster but also as an adult. Dogs in particular, engage in all sorts of endearing behaviors such as acting happy when the owner returns, wanting to be touched by the owner, touching the owner, trying to be near the owner, making the owner happy with antics, or by being obedient, looking "guilty" when they misbehave and "sad" when the owner departs. Pets can generate a feeling of well-being in people, a feeling of being loved.

Pets engage in greeting behaviors that can result in responses of affection similar to those that occur between people. The person may talk to the pet, often in the same way as to a dear friend or most likely a child. The owners may ask questions of the pet, without expecting an answer — much as one might of a small child — and might stroke, hug, or scoop up the pet into their arms.

Pets, particularly dogs, sometimes cats and even horses, greet a person in a way that is interpreted as being glad to see the person. Happy greetings tend to generate feelings of warmth in the recipient, and these feelings are of firm attachment. The greetings of pets are not the same, but the context and responses are similar to people-people involvement.

Most pets have many positive attributes that encourage attachment. Unattractive characteristics have tended, in the past, to be selected against and endearing traits actively selected for. Consequently one ends up with an individual with overwhelmingly appealing characteristics. Dogs in particular have been selected for paedomorphic morphological and behavioral traits. In fact, many breeds of dog appear to have been neotenized.

Granted, people can become quite attached to pets, but why should the attachment occur in the face of physical injury to themselves or their offspring? Certainly there are immediate benefits to keeping a pet, but these hardly seem sufficient to warrant keeping a pet that is dangerous to one's own progeny. Evolutionarily it does not make sense.

In trying to understand this, let's approach the question another way. Is there ever a time or situation when the benefits outweigh the risks of keeping something that is injurious to self or offspring? Yes. If that "something" were one's own child.

A parent does not usually get rid of a child who is a liability, especially if "getting rid of" is synonymous with euthanasia. The child is worth tolerating because the parent saves an offspring. It is even worth exposing another child to some risk if by doing so the parent is able to raise two offspring. The genetic potential of two viable offspring is greater than one; it is worth some risk of injury to one child in order to keep both offspring.

If the attachment of a person to a pet is based, consciously or subconsciously, on a parent–child relationship, corresponding risks might also be evaluated in terms of a parent–offspring relationship.

Many owners who are very attached to a pet and faced with the dilemma of separation will say, without coaxing, "But I feel toward this animal as though s/he was my child." The mechanisms of attachment have worked well to ensure parental responses and care. Women often appear to be more attached to pets, as women seem to be more easily attracted to and attached to children.

In the cases involving animals the individuals are not parent and offspring, but many of the mechanics of the relationship have been that way and the person is in an emotional trap. The emotional response is geared to enhance survival of an offspring, but in this case the recipient of the care is an alien species and the consequences are, in reality, nonproductive for the benefactor.

It is also possible that there are immediate benefits or probable immediate benefits that outweigh the costs and liabilities of keeping a dangerous pet (Smith, 1979). These immediate benefits of pet ownership may be protection, a sense of security, health, well-being, or longevity of the owner or offspring (Mugford and M'Comisky, 1975; Friedmann et al., 1979). If these benefits ultimately result in an increase of the owner's genetic representation in subsequent generations, it is worth putting up with some risks. The probability of an increase in genetic representation (benefit) must, however, outweigh the probability of a decrease in genetic representation (cost). Evolutionarily speaking, it makes sense to take risks if by doing so the probable benefits exceed probable loss. Is it possible that benefits of pet ownership really outweigh the costs of keeping somewhat dangerous pets?

Many people do make the decision to have a pet that is a liability euthanized, which is often the only available solution to what to do with the animal. But few make the decision without remorse. The mechanisms of attachment have worked well. It is rarely an easy decision.

PREVENTION OF COMMON
ANIMAL BEHAVIOR PROBLEMS

Some of the most common problems people have with their pets are based on the attachment and type of relationship they have with their pets. The most frequent complaints owners have concerning the behavior of pets are aggression related to assertion of the animal's dominance, destruction of property when the pet is left alone, and spraying or urinating and defecating in inappropriate (from the owner's point of view) places. Often such problems can be successfully treated (Tuber et al., 1974; Voith, 1975; 1977), but it is much easier to prevent their occurrence than to change them. Because these are such common problems, it would be particularly wise to take some preventive measures when pets are placed as companions with lonely people or as therapeutic agents with individuals. These people may be more likely than the general population to have animals that develop these objectionable behaviors. These people are also likely to become very attached to the animal and suffer the consequences of the behavior problem for a long time because they are afraid that the animal might be destroyed as a means of solving the behavior problem. Such individuals may also, because of a disability, be less successful in treating a problem when it occurs.

Dominance

Just as people tend to relate to pets as conspecifics, dogs tend to relate to people as conspecifics. As has been repeatedly pointed out, dogs and/or the canids from which they evolved, are pack animals who have a social hierarchy (Scott and Fuller, 1965; Fox, 1978). Varying degrees of aggressive behaviors are constantly being exhibited to test the hierarchy. A dog, particularly if socialized to people at the critical time as most pets are, relates to humans as though they are conspecifics and will relate to the human family members as pack members. The most common behavior problem of dogs is that of aggression stemming from the dog's dominant position among people.

The genetic predisposition of an individual dog, combined with how specific people related to it, often results in the dog assuming a dominant role with respect to one, several, or sometimes all family members. A dominant dog may be completely unaggressive toward strangers yet frequently threaten family members. Such a dog asserts itself in many ways. A dominant animal insists on primary access to food, sleeping places, right of way, objects of interest, or favorite individuals.

The dominant dog refuses to obey commands; it interprets commands as threats and reciprocates with a threat. An order from an owner, usually accompanied by a direct look at the dog, frequently escalates into a growling confrontation, and the owner backs down. The dog objects by growling to being disturbed, insists on right of way access to areas, resists being displaced from its resting place (which may be the bed of the owner), will not relinquish objects, often objects to prolonged petting, frequently violently objects to being hugged, and refuses to lie down on command. If the dog did obey any of these commands, it would be complying with a submissive animal's role. Instead the dominant dog objects. It does so with threats, growling, or biting, and the person backs down, thus reinforcing the dominant dog's position and further entrenching the person in a submissive role. The dog is acting perfectly normally for a dog, and its behaviors are working to its benefit.

Many dogs will jockey themselves into as dominant a position as they can and then work to maintain it. Most owners are not aware of what is happening and ever so slowly are placed into a submissive position out of which they cannot climb.

Many owners relate to the puppy then the dog as a person or child, and as a result the lines of reasoning they use in trying to cope with the situations do not work. It has been amazing to me how many people talk to the dog, discuss the situation with it, and try to use verbal reasoning to get the dog to change its behavior. Owners say, referring to the dog, "I've told him over and over again not to do that". "He just won't do what I've told him." "If you ever do that again, I'll ..."

Successful completion of obedience school does not ensure that a pet will assume or revert to a submission position in a human family hierarchy. In fact, many dogs seen in behavior specialty practices have been to obedience classes with the owner and have actually done quite well.

A social hierarchy is not maintained by only one set of acts in a specific situation but via multiple daily interactions in a variety of circumstances. A dominant animal frequently asserts itself throughout the day whenever it meets a pack member. The dominant animal stares at the other members when they meet until the submissive animals look away. The submissive animal may perform other submissive gestures such as lowering ears, head, and tail; licking the corners of the mouth of the dominant animal; moving out of the way of the dominant animal. It is because of these multiple, frequent, and often subtle encounters that obedience classes are not the sole answer to establishment or reversal of a dominance hierarchy between people and dogs.

Persons who are given pets for therapeutic reasons or as a prophylatic measure to combat loneliness may be even more likely than the general populace to find themselves playing the role of a submissive individual relative to a dog. One obvious step to combat this likelihood is to pick therapeutic dogs who assume submissive postures or behaviors toward several people. Tests such as those designed by Campbell (1973) for puppies and the Temperament Test Society (1979) for adult dogs are a start. But one should not be lulled into a false sense of security because the animal passes such tests. It is the day to day living situation and interactions between the dog and person that are going to determine over time who assumes the dominant role.

Taking precautions to ensure that a dog remains submissive to a person does not mean that an owner cannot demonstrate affection or even indulge in spoiling the pet to a degree. In fact, one can indulge a pet a great deal and still have a submissive dog—as long as one frequently insists on the dog performing certain tasks or behaviors. Owners may not have to do all of the following suggested procedures to ensure dominance of a pet but should do as many as possible.

There are several ways a placement agency could minimize the likelihood of the dog becoming dominant. In a decreasing preference order, these are —

1. Educate the owner about dominance hierarchies and dogs; give the person some procedures to do routinely and others to consistently avoid in order to prevent the dog from becoming dominant; encourage the owner to do at least a few of the procedures every day that help prevent the dog from becoming dominant.

2. Periodically visit the owner; initiate and monitor the interactions between owner and dog; have the owner perform behaviors that are interpreted as dominance gestures by the dog; ensure that the dog assumes submissive postures and performs submissive behaviors in response to the owner's activities.

3. Make frequent visits to the owner, during which a trainer takes an active role in causing the animal to assume a submissive role in the presence of the owner, trying to involve the owner in as much of the interaction as possible.

Of course, it would be best if the owner, without assistance, routinely interacted with the dog in a way that sustained the dominant status. But if an owner is unlikely to request and insure that the dog obeys commands, it may be wise to have someone visit that person and dog once or twice a week in order to practice specific procedures. The visitor could instigate or monitor small practice sessions performed by the owner. If the owner has any trouble, the visitor can help. Awareness of periodic visits of another person in a supervisory capacity may actually motivate the owners to work with the dog between sessions. If the owner is completely unable to work with the dog in any way, which is unlikely, the visitor could perform the techniques with the pet in the presence of the owner and incorporate the owner in the routines as much as possible.

**Suggested Procedures for Maintaining a Dominance Hierarchy
in Favor of a Person**

1. Never ask a dog to do something unless it is predicted the pet will obey or someone is in a position to make it obey. For example, if a dog is unlikely to come when called, it should not be called unless it is on a lead and can be pulled in.

2. Think ahead. Do not let the dog get into a situation where it is likely to misbehave and will not obey, e.g. a dog that does not come when called should not be allowed off a lead.

3. Participate in a group lesson obedience course that emphasizes rewards for good behavior. The owner is the person who takes the dog to class and trains the dog.

4. Practice obedience tasks or parlor tricks several times a week.

5. Rarely give the dog anything for free. Before the dog is fed, have it sit; before it is petted, have it sit; before the pet is allowed to crawl into a lap, have it sit immobile on the floor for a few seconds; before the dog goes outdoors, have it sit; before entering the car (if it likes to ride), have it sit; etc. Occasionally request the dog to lie down instead of sit in these situations.

6. Periodically take items away from the dog, praise the dog for being nonaggressive, then return the object.

7. Take food away from the dog, praise the dog, then return the food.

8. Periodically physically displace the dog from a resting place. Praise the dog for willingly complying.

9. Frequently restrain the dog in some way for a few minutes.

10. Frequently place hands and arms over the neck and back of the dog and exert a moderate amount of pressure. Grooming accomplishes this and could be used instead. In fact, many of these procedures could be incorporated in a grooming session.

11. Hold the dog's muzzle closed by hand for a few seconds.

12. Push the dog over on its side and hold it there for a few moments.

13. If when it is on its side the dog does not voluntarily elevate one back leg into a position of submission, lift the leg into the appropriate position.

14. Place the dog on its back, its legs in the air, immobile for a few moments.

15. Frequently look directly at the dog, until it looks away.

16. Occasionally *gently* pull the dog a few steps by the ruff of its neck or pick it up if possible.

17. Do not allow the dog to demand attention by pawing, barking, or especially by placing the forepaws on a person.

Allowing a dog to initiate a standing posture on a person is allowing the dog to assume a posture symbolic of a dominant animal and should not be allowed. The owner should not reinforce the behavior by petting or speaking nicely to the dog. The dog should either be ignored, or punished in some way (perhaps by walking away from the dog). Many owners cannot effectively punish a dog. Instead, the interaction actually becomes rewarding. The dog gets attention and the interaction becomes a game. In such cases, employing extinction is a more effective way of eliminating the behavior than using physical punishment. A technique that utilizes both extinction and punishment (isolation or separation) is to have the person simply get up and walk out of the room, totally ignoring the dog whenever it engages in inappropriate antics. Isolating the dog by removing oneself and totally ignoring the dog not only precludes reinforcement but is actually a form of nonpainful punishment that is usually effective.

The owner could also learn to anticipate in advance when the dog is likely to solicit attention. The owner then requests the dog to sit or lie down (submission postures) for attention rather than allowing the dog to demand attention by pawing or barking.

In essence, with these procedures the owner asserts him/herself as dominant in a variety of ways, several times a day. The dog is never directly reinforced for demanding behaviors; the dog is frequently asked to obey commands; the owner performs domineering gestures (requesting the dog to remain immobile, exerting pressure over the back or neck of the dog, restraining the dog, staring at the dog, taking objects away from the dog); and the dog is encouraged to assume submissive postures (tolerate restraint; look away; lie down; lateral recumbency with elevated rear leg; relinquish possession of desired objects, food, or resting places).

The more of these procedures the owner performs, the less likely the dog will become dominant. These procedures could also be repeatedly performed by a trainer before placing the animal, with the expectation that the animal's responses will generalize.

To further minimize the possibility that a pet will become dominant, female dogs could be placed as pets rather than males. More than twice as many male dogs than female dogs are presented to animal behaviorists for problems of aggression related to dominance.

Separation-anxiety/Destructive Behavior

Another of the leading problems owners have with dogs involves the attachment of the dog to the owner and appears as destruction of household items (or the house itself); excessive whining, howling, or barking; and/or eliminatory behavior when the owner is absent. Histories of such cases usually reveal a dog that has been almost constantly with the owner and rarely left alone. When the dog is left alone, it becomes anxious and engages in a variety of undesirable behaviors. Such behaviors can be treated, but it is so much easier to prevent them.

People who are given dogs for companionship and therapeutic purposes are likely to spend a great deal of time with the animal and may never leave the pet by itself. It would be wise to employ some prevention measures in order that the person can safely leave the dog when necessary.

Again, there are several approaches to working with the owner. Educate the owner concerning the potential problems of leaving a dog alone, give the owner prophylactic procedures to perform, and explain the rationale behind such procedures. The owner assumes the responsibility of performing the tasks. Alternatively, someone could visit the owner and work with the dog and owner.

Succinctly, the dog is taught to adjust to the absence of the owner by being exposed to increasing increments of absence of the owner. Within the first week the owner has acquired the dog, the owner should begin leaving the dog for very short periods of time. Initially the owner should leave the dog for only a minute or two, then return. Gradually the absence should be lengthened but not in a steady progression. Rather, the length of absence should vary with many shorter absences interjected among the longer. For example, an appropriate series of absences are 3 minutes, 5 minutes, 1 minute, 3, 1, 5, 10, 5 minutes, etc. If instead an owner were to be gone in a continuously increasing manner—1, 2, 3, 5, 7, 10 minutes, etc.—the dog can quickly learn that the owner will be gone for progressively longer periods of time. Such a dog may become more anxious. If instead the absences are less predictable in length and initially always very short, the dog comes to anticipate the owner's return at any time and soon. Hence the dog does not become anxious.

It is usually not necessary to slowly work up to several hours of absence minute by minute, particularly if the selection of the dog included the requirement that it could be left alone in a furnished house without being destructive, vocalizing, or eliminating. If the dog tolerates multiple short absences under ten minutes, the length of absence can then vary in increments of five, ten, and fifteen minutes. When the dog demonstrates that it can tolerate absences up to two hours, the pet will probably also be good for much longer periods of time.

Until the dog is gradually accustomed to a two-hour absence, it is important not to leave the dog alone for a longer time than that to which the dog has gradually been exposed. If a dog is unexpectedly left alone for many hours, it may become frightened and acquire a fearful or phobic response associated with the absence of the owner. Thereafter when this response is elicited by the departure of the owner, the owner has a problem dog.

Practice separations should be conducted at different times of the day and different days of the week. There have been cases where owners have painstakingly rehabilitated a destructive dog and could leave the dog for any length of time during the daylight hours. But to the owners' dismay, when the dog was then left for several hours at night, the dog destroyed several items. The owners had never practiced any departures at night.

It is also helpful to provide the dog with an indestructible or slowly destructible chew toy when the owner leaves. This toy may be a large rawhide toy (if the dog does not devour them too quickly), a nylon bone, or a beef shank bone. If the dog is likely to chew, these objects provide an appropriate outlet for the chewing behavior. The more attractive these objects are, the more likely the dog is to chew on them. The objects may be more attractive to the dog if they are only given to the pet when the owner leaves.

The owners should also be cautioned that if the dog is howling, barking, or crying at the door upon their return, the owners should be quiet and not enter until the dog stops the undesirable activity for at least fifteen consecutive seconds. If an owner were to enter while the dog was in the midst of engaging in undesirable behavior, the dog could associate the owner's return with the behavior. Thus, the behavior is likely to be repeated.

If the owners are unwilling or unable to comprehend the practice techniques and instructions, it would be wise to have someone periodically visit the owners and help them practice the planned departures.

Hopefully, dogs placed for therapeutic purposes have already been screened and tested for reactions to being left alone. But even if a dog has demonstrated the ability to be left alone, the pet can acquire separation-anxiety and correlated behaviors later. The danger is that the new owner may not leave the pet alone after acquiring it. Then eventually when the owner does leave, the pet responds with disturbing behaviors that escalate with subsequent departures.

The problems of aggression related to dominance and disturbing behaviors when separated from owners are two of the most common behavior problems presented to animal behavior specialists. These problems arise out of the relationship that occurs between the owner and the pet. These behaviors can occur in animals that have gone through obedience school and have performed well in the classroom situation. The problems developed because of the daily, routine ways in which the owner and pet interact.

CATS

Cat owners are not without problems either. The most common cat behavior problems presented by people are urine spraying, urinating, and/or defecating in inappropriate places. Not only do these pets stop using the litter box, but they may select items of clothing to eliminate upon. These behaviors are frequently associated with environmental events in the household.

Spraying is a method of urinating by a cat while standing with tail held straight up and quivering. The cat ejects the urine backwards onto a vertical surface. Spraying is a form of marking and communication. Intact male cats are most likely to spray but females and castrated males can also do so. Spraying is most likely to occur in environments where there are several cats; where there has been a dramatic change in the living situation such as a move, acquisition of a new pet, presence of a visitor, placement of new furniture, and/or if someone in the household does not like the cat or the cat does not like someone.

Some cats, male as well as female, react to these environmental situations by urinating in a squatting position or defecating in various locations of the house. It is interesting that cats often eliminate on items belonging to.people whom the cat does not particularly like or who do not like the cat.

When placing cats as companion animals for people, it is very important to assess the environment into which the cat is going. Not only should the cat have a temperament that is compatable or therapeutic for an individual, but the entire situation into which the cat is going should be carefully scrutinized or a behavior problem may result.

Any cat, male or female, is likely to engage in some form of urine marking if there are several cats living in the house, if there are people in the house the cat does not get along with, or if there is an obvious change in the environment. The attitude of all persons, particularly in a small household, should be positive and receptive toward cats. This may mean interviewing all members of a family before placing a cat. Subsequent introduction of animals should be done carefully and with the anticipation that urine marking may be a consequence.

In placing cats one should be aware that intact male cats are much more likely to spray than castrated males. Females are least likely to spray. Male cats also have the possibility of developing urethral obstructions, not an uncommon disorder, that must be relieved within a short period of time or the animal will become toxic and die a very uncomfortable death. If people obtaining a cat are not likely to observe whether or not the cat can urinate, it would be better to place a female cat with these individuals.

SUMMARY

There is a wide range in degree of attachment between people and companion animals. The social systems of people and a variety of domestic pets permit the human and nonhuman species to interact and reinforce affiliation for each other. Although this is an interspecies system, each species seems to relate to the other as a conspecific. For example, a pet is talked to and treated along several parameters as a person relates to another person, often as a parent would interact with his/her child.

The affiliation between people and their pets can lead to behaviors of the pet that are considered problems by the owners. Sometimes the behavior of the pet is in response to what the pet seems to interpret as undesirable behavior on the part of the person, e.g. leaving the pet alone, acquiring another pet, etc.

Neither people nor pet are acting abnormally for their own species, but when each tends to relate to the other as a conspecific, conflicts can arise.

The strength of the human–companion animal relationship is often reflected by the person's response to separation from the pet, measures taken by people to ensure that the life of the pet is "happy" and prolonged, and that the pet is not separated from the person. The cost of maintaining this relationship can be financial, estrangement from some human relationships (a friend who does not like the pet, etc.), or risk of physical injury to themselves or their children. These attitudes and behaviors of owners are revealed during the course of interviews with clients concerning their pets.

Some of the most common problems people have with pets are based on the attachment and type of relationship they have with their pet. The most frequent complaints owners have concerning the behavior of pets are aggression related to assertion of dominance, destruction of property when the pet is left alone, and spraying or urinating or defecating in inappropriate places. Dogs are often dominant toward owners and in maintaining this social position will exhibit aggression. Dogs that have rarely been separated from the owner will destroy household items, howl, cry, or eliminate when left alone. Cats often spray, urinate, or defecate in undesirable locations in response to a change in the environment or unfriendly attitudes of people or other animals in the household.

Precautionary measures can be taken to reduce the probability of such behaviors occurring. This is particularly important for people who are given pets as therapeutic agents or companions to combat loneliness.

REFERENCES

Blurton Jones, N.: Characteristics of ethological studies. In Blurton Jones, N. (Ed.): *Ethological Studies of Child Behavior.* Cambridge University Press, London, 1972.

Bowlby, John: *Attachment and Loss.* Basic Books, Inc., New York, 1969.

Campbell, W. E.: Matching puppies and people. *Dog Lovers Digest,* June, 1973.

Fox, M. W.: Socialization, environmental factors and abnormal behavioral development in animals. In Fox, M. W. (Ed.): *Abnormal Behavior in Animals.* W. E. Saunders, Philadelphia, 1968.

Friedmann, E., Katcher, A., Lynch, J., Thomas, S.: *Pet Ownership and Survival After a Coronary Heart Disease.* Presented at the 2nd Canadian Symposium on Pets and Society, Vancouver, May 30-June 1, 1979.

Keddie, K.: Pathological Mourning After the Death of a Domestic Pet. *British Journal of Psychiatry, 131*: 21-25, 1977.

Mugford, R. A. and M'Comisky, J. G.: Some recent work on the psycho-therapeutic value of cage birds with old people. In Anderson, R. S. (Ed): *Pet Animals and Society,* London, Baillicre Tindall, 1975.

Scott, J. P. and Fuller J. L.: *Genetics and the Social Behavior of the Dog.* University of Chicago Press, Chicago, 1965.

Smith, Sharon, Research Scientist, The Center on the Interactions of Animals and Society, Philadelphia, Pennsylvania, personal communication, 1979.

Temperament Test Society of America, Inc., 1677 North Alistar Avenue, Monterery Park, California, 91754, 1979.

Tuber, D., Hothersal D., and Voith, V. L.: Animal clinical psychology: a modest proposal. *American Psychologist, 29:* 762-66, 1974.

Voith, V. L. Destructive behavior in the owner's absence, Parts I and II. *Canine Practice,* 2(3): 11, 2(4): 8-12, 1975.

_____: Aggressive behavior and dominance. *Canine Practice,* 4(2): 8-15, 1977.

CHAPTER 16

PROBLEM DOGS AND PROBLEM OWNERS: THE BEHAVIOR SPECIALIST
AS AN ADJUNCT TO VETERINARY PRACTICE*

ROGER A. MUGFORD

INTRODUCTION

When a person buys or adopts a puppy, there is a clear hope and expectation that their future relationship will be a happy one. For the majority of pet owners, this expectation is realized, and a social bond is formed between man and animal to the benefit of both (Mugford, 1979). Unfortunately, however, the institution of pet ownership is not always a success. As with failures in marriage or parenthood, there are a number of reasons why the relationship between man and dog does not succeed. Feeding and care of the animal may pose an unexpectedly severe burden upon family finances, the animal may compete for time with social and business commitments of the owner, or the behavior of the animal may cause problems for the owner, the owner's family, and neighbors.

Most of these liabilities can be anticipated before acquiring the young animal. Public education programs financed by government, private agencies, and petfood manufacturers have done much to generate more realistic attitudes in would-be pet owners about the likely costs and benefits arising from having an animal in the family. However, practical advice about behavior of companion animals is difficult to distribute to a general audience. Describe the behavior of the animal and one also describes the

*The survey reported here was partly funded by the Group for the study of the Human—Companion Animal Bond. The author would like to thank Douglas Brodie, Keith Butt, Mike Findlay, and Bruce Fogle—M.s.R.C.V.S.—for tempting me away from the behavior laboratory and into practice, and Dr. Vivienne Mugford for her valuable suggestions on a draft of this chapter.

relationship with its owner. The subject is too personal and idiosyncratic to make bold generalizations about how, when, and what owners should and should not train their dog to do. But individual owners still require help and advice about their animals' behavior and effective routes to modifying behavior. To whom should they turn?

In the United States, a number of specialists in animal behavior have set up practices catering to the needs of dog owners and to a lesser extent cat owners. The "green light" for this movement came in the form of a proposal by Tuber, Hothersal, and Voith (1974), which outlined the potential of applying formal psychological principles to the task of modifying the behavior of companion animals. About a dozen Americans have answered this call, best known probably being Campbell (1975) and Tortora (1977). During the last decade there has also been an increasing interest among veterinarians in the behavioral aspect of clinical practice. Veterinarians such as Michael Fox, Benjamin Hart, and Victoria Voith have instigated this interest by their research and publications. Their influence has been greater upon the American veterinary profession than upon its counterpart in the United Kingdom. Yet the characteristics and motives for pet ownership are broadly similar in both these countries, and pet owners probably encounter the same types of animal behavioral problems.

I am a comparative psychologist who has studied the behavior of dogs and cats and their interactions with people over many years. Following the formation in 1979 of an international organization to promote research and understanding of the human–companion animal bond (Group for the Study of the Human–Companion Animal Bond), I decided to investigate the types of behavioral problems that worried British pet owners and where possible gauge the relative frequencies of behavioral problems in various sectors of the total population. The most practical mechanism for conducting this work was to establish a behavioral practice to which problem pets or problem owners could be referred by veterinarians for treatment.

A major objective of this practice is to critically evaluate the efficacy of alternative approaches to treatment of animal behavioral disorders. A plethora of treatment options for identical behavioral conditions has been advocated by past authors, whether writing for a lay or a scientific/veterinary readership. Unfortunately, few of these publications have attempted to quantify the long-term impact of their chosen treatment strategy upon the animal and its owner. Most present a series of case histories to support generalizations about behavior and methods to eliminate particular behavioral problems.

This chapter is in the nature of a progress report on the first nine months of a continuing field study of the human–companion animal bond. The most important two findings to emerge were (1) over three-quarters of all serious behavioral problems in dogs could be satisfactorily resolved, and (2) veterinary services such as drug administration, neurological examination, or gonadectomy were required to support the behavioral objectives in over 50 percent of the cases. Thus, the psychologist or dog trainer who works in this area without close collaboration from the veterinary profession is at a considerable disadvantage to one who is part of a multiprofessional team.

THE STUDY POPULATION

Between June 1979 and March 1980, 125 dog owners and 5 cat owners were referred to the author by 58 veterinarians. The relative paucity of cat cases says something about either cats or cat owners, but this chapter will only deal with behavioral problems encountered by the author in dogs. Of the dog owners, 70 percent lived in the London or Greater London region, and only 16 percent were from rural communities. While this breakdown between city and country could point to the urban environment as a predisposing factor for behavioral problems in dogs, it is equally likely to reflect the nature of small animal veterinary practice in the United Kingdom. Thus, exclusively small animal veterinarians may be more likely to respond to their clients' request for help about a behavioral problem than veterinarians working in mixed practice.

The grounds for referral generally concerned the behavior of the dog, but 12 percent of the cases were referred because the veterinarian was also concerned about the unrealistic attitudes and expectations of the owner. Of the owners, 40 percent were referred because they had asked their veterinarians to euthanize or neuter their animal, and it was only discussion arising from this request that highlighted the presence of a behavioral problem. Most owners were surprised and perhaps a little self-conscious to be referred to a behavior specialist, who was often labelled as a "dog psychiatrist" or "shrink," who was expected to come armed with a mobile couch and personality inventory for maladjusted canines! More seriously, most owners had not expected their veterinarians to be so interested or helpful about the problems that they were experiencing with their dogs' behavior. This is obviously a point of public ignorance about which the British veterinary profession might concern itself in the future.

APPROACH TO TREATMENT

Each client was visited at home by the author. So far as possible the whole family was involved in a discussion about the dog's behavior, and this often required that visits be made on weekends or evenings. The self-imposed rule that one always saw the animal and owner in their home environment posed considerable time and cost penalties, and the alternative of working from an office, clinic, or training center seems to be an attractive one. Behavior specialists doing similar work in other countries have generally opted to have pet owners come to them (*see* Chapters 12 and 15). Victoria Voith and colleagues at the University of Pennsylvania Veterinary School have owners complete a questionnaire. This and further discussion at the clinic is then used to guide treatment.

My experience suggests that no amount of questioning can identify some of the more subtle social interactions that occur between the human and animal members of a family; and a certain voice, posture, or touch can be the critical but unspoken factor that reinforces an undesired behavior pattern. The owners are unlikely to be aware of the salience of these cues, which are also unlikely to be revealed to an observer in a strange environment (Mugford, 1980). Another sound reason for home visits is that one is able to incorporate the existing layout of rooms and furniture into the therapeutic regime and to build upon the traditions and expectations of the household. The alternative is to recommend a stereotyped regime that may not be in tune with the family's needs. It is my impression that a home visit played an absolutely essential part in resolving some fifty percent of the cases, and most of the remainder benefited from a home visit. There is no data-base presently available that allows one to compare the efficacy of treatments with and without a home visit; that is something for the future.

About two hours were spent with each client, and the chosen technique(s) was outlined, demonstrated, and rehearsed with the whole family. The rehearsal stage was particularly important because many dog owners required considerable coaching in their vocal and physical control of the dog. Following the visit, each client was sent a typed report with a copy to the referring veterinarian (*see* Appendix 16-1). The objective of this report was twofold: it reiterated advice given earlier and committed the owner to make a conscientious effort to resolve the problem. Clients were always invited to maintain contact by correspondence or telephone, and second or subsequent visits were made at no additional fee. Most clients appreciated having this "open line" to a behavior specialist, as did collaborating veterinary surgeons. At the time of writing (April 1980), the fee was £30 (70 dollars), which is about the same as the cost of having a bitch spayed in the London area.

CHARACTERISTICS OF PROBLEM DOG OWNERS

The description given by Voith (*see* Chapter 15) of the owners of dogs exhibiting behavior problems in the United States is matched by my experience in the United Kingdom. Clients were referred from all age groups and socioeconomic backgrounds. No particular psychological trait could be seen that distinguished people referred to me as clients from any other group of "normal" pet owners, and they were certainly not all excessively soft or indulgent toward their dogs. Like Voith's sample, many had taken their dogs to training classes and achieved a fair standard of obedience control. Only 36 percent of the clients had poor obedience control over their dogs; 30 percent had trained them to a high standard, and 34 percent could make them SIT/STAY/COME and HEEL. It seems that the techniques and authority obtained during standard obedience training did not necessarily transfer to the situation at home where most behavioral problems were encountered. Indeed, four of the clients had each spent more than £100 having their dogs professionally trained in obedience work, and one client was himself a dog trainer.

Clients did, overall, tend to have above average income, but this is to be expected among any social group paying for a professional service. Thus, 52 percent of all clients were in socioeconomic class A and B, whereas the wider British dog-owning population incorporates only 22 percent of social class AB (Pedigree Petfoods, 1980). This does not mean that rich people are liable to own problem prone dogs but rather that my sample was entirely drawn from dog owners who have consulted veterinary surgeons.

The most significant psychological attribute of the owners of dogs with behavioral problems was that they had a strong attachment and committment to the animal. Only two of the 125 clients could be said to have had a careless or detached view of their dogs; the remainder loved their animals despite the problems and risks that they presented.

TYPES OF BEHAVIOR PROBLEMS

Each of the 125 cases represented unique qualities of a truly social relationship between animal and man. The majority of the behavioral problems for which owners sought help were essentially problems of relationships, by which I mean that the dog was only a problem in the context of its home and owner. Put the dog in a new environment and its behavior problem may not be immediately apparent.

The idiosyncratic nature of behavioral problems makes their classification difficult and even misleading, but I have defined each class before going on to discuss its relative frequency in the sample of 125 cases.

Aggression

Aggression is the act or the threat of injury to one animal by another. It is not a unitary concept and the biological basis of the various types of aggression are not neccessarily related to one another (Moyer, 1968). Houpt (1979) suggests six categories of aggressive behavior for the dog, namely territorial, predatory, sexual, maternal, irritable, and fearful aggression. The classificatory schemes of both these authors tend to emphasize the state or intent of the animal that is initiating an attack. In my experience, it is more realistic to consider the target or victim when classifying the types of aggressive behavior in animals (Mugford, 1973). Thus, I have divided the aggression cases as follows: (a) where the owner or family members were threatened by the dog, (b) aggression toward visitors and strangers, and (c) aggression toward dogs. Predation by dogs against members of another species, such as sheep, cats or poultry, is excluded from this consideration of aggression because it relates more closely to play and ingestive behavior (Polski, 1975).

Aggression toward owner (20% of the cases)

The owner and members of the owner's family often became victims of bites and threats from the overly dominant dog. Such attacks occurred within the normal context of canine social organization, although there may have been predisposing genetic factors on the part of the dog and personality traits on the part of the victim. These cases responded very well to a combination of the behavioral techniques outlined by Voith (*see* Chapter 15) coupled with hormonal treatments.

However, the owner and others well known to the dog were sometimes bitten in situations where there was no apparent provocation or point of competition. Indeed, attacks occurred with no warning growl or threat from the dog; just before, it may have been playing or asleep. In forming a differential diagnosis between a normal canine dominance problem and cases where aggression is quite unpredictable, I find it helpful to mention the R. L. Stevenson characters Dr. Jeckyl and Mr. Hyde. Dogs that fit the latter profile can be extremely dangerous, and owners should not challenge the dog in a dominant fashion. The condition has been very well described by Van der Velden, De Weerdt et al. in the Dutch sennenhund breed (1976). These authors demonstrated a marked genetic predisposition of certain breeding lines of sennenhunds, but they were unable to find clinical, neurological, or endocrinological abnormalities that could account for the behavioral disturbances.

I have encountered the "Jeckyl and Hyde" syndrome in six cocker spaniels (five red cockers), three Old English sheepdogs, one sheltie, and two dogs of mixed breed. Such dogs can be heartbreaking to own because they may be placid and lovable for 90 percent of the time but become monsters thereafter. Some of these cases have responded well to dietary manipulation and anticonvulsants, but the prognosis is generally poor. Interestingly, these dogs are unlikely to be spotted in a veterinary examination room or the pedigree dog show ring because the symptoms are suppressed in novel surroundings. A genetic selection program to eliminate this trait would, therefore, have to rely upon the honest evaluation of the owner/breeder.

Aggression toward visitors and strangers (18% of the cases)

This is the problem known to mailmen the world over because the dog is most aggressive within and at the margins of its home area. Away from home the dog is often friendly and approachable to strangers. A most effective behavioral treatment involves coaching the owner to punish early expressions of aggression (rather than to hold or pet the dog) and coaching visitors to be passive and offer food rather than challenge the dog. The best protection for mailmen against such dogs is, in my experience, a packet of palatable semimoist dog food rather than ultrasonic alarms, sticks, pepper bombs, etc.

Aggression toward dogs (14% of the cases)

Many urban dogs make most of their social contact with other dogs from the protected perspective of their leash and often become highly aggressive in the process. The response can be highly stereotyped and unrelated to the sex of the "victim." Although the problem is more commonly encountered in male dogs, it does not neccessarily respond to castration. Treatment of this condition usually involves three phases: oral dosing with megestrol acetate,*

---

*Ovarid: Glaxovet, Greenford, Middlesex, United Kingdom.

graded exposure to placid "stooge," dogs and behavioral reorientation of the owners so that they can punish early expressions of aggression but reward tolerance.

Toilet Problems (15% of the cases)

There is an almost universal taboo among dog owners against their animal eliminating indoors. A surprisingly large number of adult dogs were referred because they urinated, defecated, or did both in the owner's home. Of this group, two-thirds were males, where the problem was often urine marking against vertical objects. Treatment of the latter cases usually centered upon hormone therapy. Urination in a nonsocial context and defecation indoors occurred in all sexes but seemed to be especially common in dogs living in large townhouses with small gardens. Treatment of these house-soiling dogs involved application of "first principles" of toilet training (such as environmental restriction), strictly scheduled meals and exercise, and reward for elimination in desired locations.

Isolation Anxiety (23% of the cases)

The dog that became anxious when left alone created a number of related problems for its owner: destructiveness (12% of the cases), barking (9%), and urinating or defecating (3%). Considerable expense and inconvenience was borne by the owners of such dogs, as noted by Voith (*see* Chapter 15). Because these dogs have a strong social attachment to their owner, they are usually much loved despite the cost. It is my experience that owners are conscientious in instituting a corrective training strategy to make the dog less dependant upon their company. The prognosis is, therefore, good.

Automobile Related (6% of the cases)

Personal observation suggests that many dogs make rather tiresome travelling companions. While only a small number of such cases (8 out of 125) were presented for treatment to the author, I suspect that it is a common problem. Only a minority of dogs became anxious or fearful when travelling in an automobile (two cases seen). It was more common to encounter the dog that was hysterically overexcited at the onset of a journey. This seemed to be largely an anticipatory response to the pleasant experiences that followed an automobile journey, namely off-leash excercise, contact with other dogs, or simply a change of environment. The dog's howling and general boisterousness was sometimes sufficient to impair the owner's driving skills and thereby imperil other road users.

Tranquilizers are usually not an appropriate form of medication for the hysterical "superenjoyment" traveller. Behavioral modification can be successful where it is designed to remove the post journey reinforcement, e.g. walks. Mock departures and journeys " 'round the block" were usually suggested in order to reverse the reinforcement contingencies that originally created the problem. As with many other behavioral traits, prevention by counselling puppy owners is preferable to reversing an established habit of the adult dog.

Neurological and Physiological Disorders (14% of the cases)

There were eighteen dogs presented for evaluation and treatment whose behavior was qualitatively outside the canine norms and for which no external or owner-related causes could be found. My use of the term "neurological" is open to criticism, especially considering the scant availability of clinical neurophysiological skills and facilities in the United Kingdom. Only two dogs were investigated by a veterinary neurologist, so the remaining sixteen dogs were classified by subjective rather than clinical criteria.

Many of the dogs in this category exhibited "Jeckyl and Hyde" character changes as described in an earlier section of this chapter. Other cases exhibited behavioral problems associated with clinical epilepsy, and the first line of treatment was obviously to stabilize the epileptic condition with anticonvulsants.

There were four cases referred with behavioral problems that stemmed from a previously undetected physical disorder: middle ear and skin infections and inflamed anal sacs. These cases again emphasize the importance of subjecting the dog to a careful physical examination by a veterinarian before delving too deeply into its behavioral condition.

Obedience (29% of the cases)

Although one-third of the dog cases posed problems for their owners because of a lack of obedience control, it is more remarkable that the remainder did not pose obedience problems. Clients who had attended dog training classes were generally not impressed by the facilities provided or the methods employed by trainers. The most frequent complaint about dog trainers was that they placed too much emphasis upon harsh correction of minor faults, in both dog and owner alike.

Unfortunately, the methods and philosophy of many dog trainers in the United Kingdom seem largely to be rooted in the military traditions of World War I. There is considerable scope for developing a more enlightened and humane approach to training pet dogs, the objective being to employ regimes that build upon the natural response tendencies of both animal and owner. For instance, every psychologist knows that food is a powerful reinforcer for learning in animals, yet most dog trainers abhor its use.

Other (19% of the cases)

Dogs can acquire a seemingly limitless variety of eccentric behavior patterns from their owners, and many of the problems I dealt with defy classification. A common theme of acquired neurosis was that the dog engaged in certain idiosyncratic behaviors in order to obtain further social contact with its owner. Examples are a dog that scratched the carpet before feeding, another that licked furniture and walls in the owner's presence, a cavalier that gazed at shadows, and a dachshund that hoarded cushions in an obsessive manner. All these ruses enabled the dogs to obtain operant control over their owners' behavior: clever canine psychologists! A very high level of success in overcoming such attention-seeking behavior patterns can be obtained by explaining the principles of learning theory to the owner and having him reverse the reinforcement contingencies that created the problem in the first place.

Other behavioral conditions that were infrequently encountered were as follows: coprophagia (three cases), stock chasing (five cases), sexual mounting (four cases), and various phobias (four cases). With the exception of the latter, these are essentially normal canine behavior patterns that pose problems to their owners because emotional or practical considerations make them undesirable in a pet dog. The dog that eats feces or chases sheep faces a high risk of euthanasia, but the conditioned-aversion technique outlined by Garcia, McGowan, et al. (1972) and Mugford (1974) offers a promising line of treatment for such cases.

RELATIVE FREQUENCY OF BEHAVIOR PROBLEMS

The 125 dogs referred for treatment by the author exhibited a total of 199 behavior problems. Of course, one could argue the definition of what constitutes a behavior problem, but my criterion centers upon whether or not the dog met the *reasonable* expectations of the owner. About one-third of the sample posed more than one problem, the overall average being 1.6 problems per dog. Some of these multiple problems were clearly related, e.g. the aggressive dog suffering from a neuropathological condition, whereas many others were not related, e.g. toilet problems and poor obedience.

A summary of the main findings appears in Table 16-I. The most commonly encountered problem was aggression in its various forms, which appeared in more than half of the dogs. After aggression, the rank order of frequency was obedience problems, isolation anxiety, toilet problems, problems arising from various neurological and metabolic disorders, automobile related problems, and a substantial assortment of miscellaneous problems (19% of the total sample).

Sex of Dog

An analysis of the results by sex of the dog is shown in Table 16-II. Of the animals referred, 60 percent were intact male dogs, and they were significantly overrepresented in all of the problem categories. The small numbers of castrated male dogs with behavior problems does not necessarily mean that castration is a useful

TABLE 16-I

INCIDENCE OF VARIOUS BEHAVIORAL PROBLEMS
IN DOGS REFERRED FOR TREATMENT

Behavior problem	% dogs exhibiting problem
Aggression to owners	20
Aggression to visitors	18
Aggression to dogs	14
Toilet related	15
Isolation anxiety	23
Automobile related	6
Neurological	14
Obedience related	29
Other	19

NOTE: the sum of the percentages is > 100 because of multiple problems in some dogs
(see text). N = 125.

TABLE 16-II

INCIDENCE OF BEHAVIORAL PROBLEMS BY SEX OF DOG

Behavior problem	Intact male N = 74		Castrated male N = 15		Intact female N = 23		Spayed female N = 13	
	No. cases	%	No. cases	%	No. cases	%	No. cases	%
Agg. to owners	19	16	3	13	3	9	0	0
Agg. to visitors	10	8	2	9	5	14	5	24
Agg. to dogs	12	10	1	4	3	9	2	10
Toilet	12	10	1	4	5	14	1	5
Isolation anxiety	17	14	2	9	7	20	3	14
Automobile related	5	4	0	0	0	0	3	14
Neurological	9	7	4	17	5	14	0	0
Obedience related	21	18	5	22	6	17	4	19
Other	15	13	5	22	1	3	3	14
Totals	120	100	23	100	35	100	21	100

therapeutic procedure. Rather, it is my impression that the number of castrates in the total United Kingdom dog population is small (no data available), and cases were often referred to me after castration had failed to achieve the declared behavioral objectives. This view is consistent with the results obtained by Hopkins, Schubert, and Hart (1976) in a retrospective survey of the behavioral effects of castration upon pet dogs; in general, castration is not as effective in altering canine behavior as it is in altering feline behavior.

Only 28 percent of the referred sample were female, of which about one-third were spayed bitches. Bitches comprise about 47 percent of the total United Kingdom dog population (Goodwin, 1975), so they were significantly underrepresented in this sample of problem dogs. However, the grounds for referral were very similar in bitches and dogs. Interestingly, aggression toward the owner, visitors, or other dogs accounted for 31 percent of all behavior problems in intact bitches and 34 percent in intact males, which emphasizes that aggression is not a male prerogative.

Breed of Dog

Table 16-III provides a breakdown of the case load by breed of dog. According to Pedigree Petfoods (1980), 54 percent of the dogs in the United Kingdom are purebred or pedigree, and the rest are crossbred. It may, therefore, be significant that only 12 percent of my referral sample were crossbred, and as a group they would seem to be less prone to developing behavioral problems than purebred dogs. Of course there may be a number of sociological and economic parameters that distinguish the owners of purebred dogs from those owning crossbred dogs, but hybridization between breeds should also be considered as a potentially positive factor in programs intended to reduce the incidence of behavioral problems in dogs. I do not expect this suggestion to be well received by those who show and breed pedigree dogs, but the genetic selection criteria used in these circles are almost entirely oriented toward visual and physical characteristics. Selection for behavior and temperament receives remarkably little systematic attention by dog breeders, yet this is probably the main criterion of "fitness" in the pet dog.

TABLE 16-III

BREEDS OF DOG REFERRED WITH A BEHAVIOR PROBLEM

Breed of dog	Number referred	% of all cases
Labrador	11	9
Cocker Spaniel	10	8
German Shepherd	10	8
Spaniels (Springer & American)	8	7
Dachshund	7	6
Cavalier K. C.	6	5
Collie (Border & Rough)	5	4
Doberman	5	4
Poodle (Toy & Miniature)	5	4
Boxer	4	3
Yorkshire Terrier	4	3
Afghan	3	2
Cairn	3	2
Golden Retriever	3	2
Old English Sheepdog	3	2
Other purebred	23	19
Crossbred	15	12
Totals	125	100

There is a strong heritability component to canine behavior (Pfaffenburger, Scott et al., 1976), so it came as no surprise to find that individual breeds were particularly prone to develop certain behavioral conditions. The most marked example of this phenomenon was the cocker spaniel, with a high incidence of individuals making unprovoked attacks upon their owners. Cases involving dobermans and German shepherds often conformed to the stereotype for these breeds, i.e. dominance related problems, while dachshunds seemed to be particularly prone to aggression against other dogs. Of the seven dachshund cases, five were pairs of dogs that had not developed a stable dominance relationship with one another and so fought regularly.

The breeds of dogs that were not referred to the author are worth a mention: corgis, setters and beagles were surprisingly absent, while poodles and West Highland terriers were under-represented compared to their numbers in the United Kingdom dog population.

Age of Dog

The mean age at which dogs were referred was 2.9 years, (standard deviation, 2.3 years.) There were 65 percent that were less than two years old but only 11 percent were less than one year old, which suggests that behavior problems in dogs are mostly a feature of the young adult. However, cases were seen in dogs of all age groups, from fifteen weeks to fourteen years. In most of the cases, the owners reported that they had first noticed the undesired behavior pattern when their dog was about a year old.

There is a suggestion in the data that dogs with behavior problems are more likely to be acquired at a later age than dogs that do not pose a problem to their owners. Of the sample, 28 percent had been reared by a previous owner, but then it had either been rejected or placed in a new home via an animal rescue agency. Unfortunately, no comparable data exists on the age of acquisition of all dogs in the United Kingdom. Some dogs did seem to be particularly liable to lose the affection of their owners, and one case had been placed in new homes on four occasions. Wilbur (1976) noted that behavior problems were given as the reason why a dog was "given away" by 21 percent of former American pet owners. Beck (1980) considers that dogs exhibiting behavioral problems run a particularly high risk of being rejected by their owners and thereafter entering the stray population. The available evidence certainly suggests that welfare and statutory agencies interested in tackling the problem of free roaming and unwanted dogs at source could well direct their energies toward helping existing owners to avoid or to resolve behavioral problems in their dogs.

RESULTS OF TREATMENT FOR BEHAVIOR PROBLEMS

After three months of treatment, clients were sent a questionnaire to evaluate progress and to remind them to seek further help if it was needed (*see* Appendix 16-2). Of these follow-up questionnaires, sixty-four were dispatched, of which twenty-seven have been returned at the time of writing. The full results of this continuing survey will be reported elsewhere, but replies to questions 4, 6, 7, and 9 will be considered here, albeit based upon a small number of returns.

Table 16-IV indicates that all respondents considered that their dog posed a serious or a very serious problem at the time they were referred to me. Indeed, 33 percent of the owners were considering having their dogs euthanized.

The effects of treatment upon the dogs' behavior are summarized in Table 16-V. The success rate as gauged by whether the

TABLE 16-IV

ATTITUDES OF OWNER TO BEHAVIOR OF DOG*

Question 4:	How serious was the worst problem? N = 27	
		%
	Very serious	56
	Serious	44
	Not serious	0
Question 9:	Would you have considered euthanasia? N = 27	
		%
	Yes	33
	No	67

*See posttreatment survey, Appendix 2.

behavior had "greatly improved" or "improved" was high: 78 percent. The residual 22 percent mostly comprised the neurological category, as was described earlier, but inevitably there were failures to resolve other types of behavioral problems. The reasons for failure are obviously very important for designing more effective treatment strategies in the future. Was inappropriate advice given or was there a lack of commitment from the owner?

TABLE 16-V

RESPONSE TO TREATMENT*

Question 6: Did the behavior change following treatment? N = 27	
	%
Greatly improved	19
Improved	59
No change	11
Got worse	11
Question 7: If improved, how long to take effect? N = 21	
	%
Immediately	48
1–3 weeks	33
4 weeks	19

*See posttreatment survey, Appendix 2.

If there was to be an improvement in the dog's behavior as a result of treatment, it tended to occur rather quickly. Remarkably, half of the cases were said to improve immediately following my visit, a third within one month, and the rest took longer to respond.

One most important finding that emerged from working in this area has been the willingness of practising veterinarians to refer clients to a specialist in animal behavior and thereafter to collaborate in designing effective treatment strategies. Of the cases studied, 54 percent required an active and continuing involvement of the referring veterinarian before they were satisfactorily resolved. Apart from the beneficial effects of being able to share the results of a clinical examination of the dog with myself, the veterinarian also had a crucial role in prescribing behaviorally active pharmaceuticals such as progestins, performing a surgical procedure such as a gonadectomy, or by initiating clinical tests that could shed additional light upon the behavioral condition of the animal.

CONCLUDING REMARKS

It is still a surprise for me, a psychologist, to encounter the common belief that human behavior is predestined. But if one accepts that this is a commonly held belief, then it is not surprising that many dog owners are sceptical about the ability of a veterinarian or animal behaviorist to modify their pets' behavior in a positive direction.

In fact, we have seen that three out of four serious or very serious problems in dogs can be satisfactorily resolved by applying a combination of veterinary and behavioral techniques. The relationship between man and his or her "best" friend can be preserved even when it has appeared to the owner that only euthanasia provides the remedy for an irritating or dangerous behavioral trait in the dog.

Despite these successes in helping individual clients, the first nine months of operating a behavioral referral service has revealed more scientific problems and areas of technical ignorance than I had expected. For instance, why are red cocker spaniels so prone to the "Jeckyl and Hyde" syndrome, and why do individual dogs respond so differently to progestins, to dietary manipulation, even to castration or spaying?

Apart from the obvious requirement for more research into these questions, there is also a need for programs that can reduce the incidence and seriousness of behavioral problems in dogs. Existing scientific information about behavior genetics and the prevention of behavioral problems in animals could well be disseminated to a wider audience than at present. The initial target audience should, perhaps, be veterinarians, breeders, show judges, and trainers: those who can influence attitudes of the public toward companion animals.

APPENDIX 16-1

Mr. and Mrs. F. Jones
29, Belton Avenue
London, SW7

Dear Mr. and Mrs. Jones,

Re: Buttons

Following my visit on May 3rd, I am writing to reiterate advice on Button's behavior. The problems that he poses fall into three categories: toileting indoors, anxiety when left alone, and general obedience. I think that the first and second may be partly related, i.e., he is more likely to toilet at night in the kitchen because he is anxious. However, I will deal with them as separate topics as follows:

1. The enclosed leaflet on toilet training is mostly relevant to your situation. In particular:
   a. Close confinement at night to about 1 square metre, then gradually increase the floor area available as he "notches up" clean nights.
   b. Feed him as early in the day as possible.
   c. Increase his exercise to say, a walk on the streets or parks first thing, at 6:30, and late at night.
   d. Prevent access to upstairs by a gate, or better by training him to regard it as a no-go area.

   If you have not succeeded in stopping his leg cocking indoors within one month, then you should consult Mr. Butt about a veterinary treatment that will temporarily suppress male behavioral patterns.

2. The other leaflet refers to treatment of dogs that are destructive when left alone. While Buttons is not destructive, he does become over anxious when out of your company, and the same approach is needed to loosen his social dependence upon the Gainsford family. In particular, I recommend that you cut down random petting and cuddling of Buttons: try to restrict this form of reward to occasions when he is being good. When

you return to visit him in a room where he is calm and alone, make your greeting very effusive. Likewise reward tolerance of being alone in the house, but try not to reward the anxious response.

3.   Do persevere with standard obedience training: particularly to stop leash pulling and develop an "electric" COME! response. Enforce the latter using the "Flexi" lead. The door scratching habit could become a serious problem for you, so punish him for it, wait for 5 minutes, then let him in on the cue of quietness or nonscratching.

Do contact me again if you require further help or advice: it will be gladly given at no additional fee.

Yours sincerely,

Dr. R. A. Mugford.
c.c. Mr. K. Butt, M.R.C.V.S.

APPENDIX 16-2

QUESTIONNAIRE USED FOR 3-MONTHS FOLLOW-UP OF OWNERS OF PROBLEM DOGS

Your Name _____   Your dog's name _____

1. Please list the good and the bad aspects of your dog's behavior in their order of importance to you. For example, a loving nature might be a good point and incessant barking a bad point.

    Good Points   *Bad Points*

    _____   _____

    _____   _____

    _____   _____

2. Which of the above "bad points" was the problem for which you sought help?

    _____

3. Apart from the veterinary surgeon and Roger Mugford, had you sought advice from any other source about your dog's behavior?

    Answer yes or no _____

    If yes, who did you ask for advice? _____

4. At the time Roger Mugford visited, how much of a problem was the dog's behavior to you? Tick one of the boxes below.

    ☐ very serious   ☐ serious   ☐ not a serious problem

5. Did you follow Roger Mugford's advice? Please tick one of the boxes below.

    ☐ totally   ☐ in part   ☐ not at all

6. Following Roger Mugford's visit, has the dog's behavior changed? Please tick one of the boxes below.

    ☐ greatly improved   ☐ improved   ☐ no change   ☐ got worse

7. If there was an improvement in your dog's behavior, how long did it take for the treatment recommended by Roger Mugford to have an observable effect? State number of days or weeks before you could see an observable effect _____

8. How long did the improvement, if any, continue? _____

9. If your vet had not recommended you to see Roger Mugford, what other steps might you have taken to overcome your dog's behavior problem? _____

    _____

    Would you have considered having it euthanased (i.e. put down)? _____

10. Do you have any other comments (a) about the present behavior of your dog or (b) about the service provided by Dr. Mugford? Continue overleaf if more space is required _____

    _____

    _____

REFERENCES

Beck, Alan: Director, The Center on the Interactions of Animals and Society, Philadelphia, personal communication, 1980.

Campbell, W. E.: *Behavior Problems in Dogs*. American Veterinary Publications, Inc, Santa Barbara, 1975.

Evans, J.: Current thoughts concerning the control of hypersexuality in dogs, with particular reference to the role of progestagens. In *Refresher Course in Community Medicine, Proceeding no. 37*. University of Sydney Post-Graduate Communications in Veterinary Sciences, 1978.

Garcia, J., McGowan, B. K., and Green, K. F.: Biological constraints on conditioning. In Black, A. and Prokas, W. F. (Eds.): *Classical Conditioning*. Appleton-Century-Crofts, New York, 1972.

Goodwin, R. D.: Trends in the ownership of domestic pets in Great Britain. In Anderson, R. S. (Ed.): *Pet Animals and Society*. Bailliere Tindall, London, 1975.

Hopkins, S. G., Schubert, T. A., and Hart, B. L.: Castration of adult male dogs: effects on roaming, aggression, urine marking and mounting. *JAVMA, 168*: 1108–10, 1976.

Houpt, K. A.: Aggression in dogs. *The Compendium on Continuing Education for the Small Animal Practitioner*, 1: 123–28, 1979.

Moyer, K. E., Kinds of aggression and their physiological basis. *Communications in Behavioral Biology*, 2: 65–87. 1968.

Mugford, R. A., Androgenic stimulation of aggression—eliciting cases in adult opponent mice castrated at birth, weaning or maturity. *Hormones and Behavior, 5*: 93–102. 1974.

_____ : The social significance of pet ownership. In Corson, S. A. (Ed.): *Ethology and Non-verbal Communication in Mental Health*. Pergamon Press, New York. 1979.

_____ : The owner as a social reinforcer of abnormal behavior in dogs. *Group for the Study of the Human–Companion Animal Bond Newsletter*, 1: 9–11, 1980.

Pedigree Petfoods: *Pets and the British*. Melton Mowbray, Leics., 1980.

Pfaffenberger, C. J., Scott, J. P., Fuller, J. L., Ginsburg, S. E., and Bielfelt, S. W.: *Guide Dogs for the Blind: Their Selection, Development and Training*. Elsevier North Holland Inc., New York, 1976.

Polski, R. H.: Development factors in mammalian predation. *Behavioral Biology, 15*: 353–82, 1975.

Tortora, D.: *Help, This Animal is Driving me Crazy*. Playboy Press, Chicago, 1977.

Tuber, D. S., Hothersal, D., and Voith, V. L.: Animal clinical psychology: a modest proposal. *American Psychologist, 29*: 762–66, 1974.

Van der Velden, N. A., de Weerdt, C. J., Broojmans-Schallenberg, J. H. C., and Tielen, A. M.: An abnormal behavioural trait in Bernese Mountain Dogs (Berner Sennenhund). *Tijdschr Diergeneneesk*, 101: 403–7, 1976.

Wilbur, R. H.: Pet ownership and animal control, social and psychological attitudes, 1975. *Report to the National Conference on Dog and Cat Control, Denver, Colorado*. 1976.

CHAPTER 17

CLIENT PROBLEMS AS PRESENTED TO THE PRACTICING VETERINARIAN

ANDREW YOXALL

This review discusses aspects of the human–companion animal bond presented in the veterinarian's reception and examining rooms. Because of the very nature of the topic the material is anecdotal, and pet ownership categories are described in practical terms as the veterinarian sees them.

Problems in the human–companion animal bond can arise directly from the interaction between pet and owners. Personal, psychological, or social problems of the owner can also manifest themselves in the relationship with the pet. In addition, the pet can be passively used by the owner for the expression of these problems.

DIRECT INTERACTION PROBLEMS

Direct interaction problems can occur because of inappropriate behavior in the animal or the owner's failure to establish a satisfactory relationship with the animal. The majority of cases have elements of both. Until very recently most veterinarians had no professional training in animal behavior, let alone human behavior. However, understanding human behavior as well as animal behavior is of increasing importance to the practice of companion animal veterinary medicine because of the growing importance of the companion animal within our changing society and the greater variety of pet–owner interactions that can occur.

The nature of a particular human–companion animal bond is affected greatly by the basis on which animal choice is made. When the purchase of a dog is contemplated, the prospective owner has a preconceived notion of how the dog will behave and how it will fit into the existing family. The breed of dog chosen is likely to be determined by the owner's conception of how the breed type behaves as well as by other factors, such as the color and shape of the animal. Although the behavioral patterns of the newly acquired puppy will be molded to some extent by attitudes of the owners, the animal's intrinsic characteristics will remain, and so anomalies may arise if there is conflict within a family. The wife's choice, if she is decision maker, may be a small dog, yet the husband may really have wanted an Alsatian and may attempt to imbue the dog with at least some of the characteristics of the bigger breed.

Many owners demand that a dog act as a watchdog or guard and at the same time express only friendly behavior toward individuals identifiable to the owner as friends. Although many dogs are well able to fulfill such a dual role without any problems, it is tempting to attribute some of the behavioral problems exhibited by pet dogs to inability to reconcile these ambivalent demands.

There is very little investigation of whether or not owner attitudes can be correlated positively with pet animal behavior, but at a purely subjective level the practising veterinarian soon learns that certain types of owners are likely to have dogs of certain temperaments.

The young owner entering the clinic in a defensive manner, with a similarly defensive young Alsatian or Samoyed encourages a cautious approach to the animal by the veterinarian and his lay staff. In one such case encountered by the author, the owner admitted that the dog had already bitten several people. It transpired that he actually approved of this behavior. He claimed that this demonstrated that the dog was capable of protecting his wife and children, although one might be tempted to suppose that in fact this particular individual was gratified to see *his* dog expressing the aggressive instincts present in himself. If the need to dominate is accepted as one of the basic needs of any individual (Argyle, 1967), then one might anticipate displacement of this sort. This problem resulted from the interaction of behavioral characteristics of the dog *and* of the owner.

In another contrasting case, the attitudes and characteristics of the owners were very much more important than the characteristics of the dog, which was in many ways being "used." An Alsatian was presented for euthanasia because it was "biting people." On further questioning, it transpired that the dog had not actually bitten anyone but showed threatening behavior toward certain individuals. Further questioning revealed that the owner's wife was particularly dominant and strongly resented the close relationship that existed between the husband and his dog. She used the supposed display of aggression as an excuse to remove the dog.

Cases in which problems relate solely to the dog may be difficult to define but are perhaps seen best in certain dogs acquired from animal shelters or homes for unwanted animals. Many of the animals in these places have been rejected because of inappropriate behavior, and the new owner may be presented with quite considerable behavioral problems.

Cases of direct interaction problems may fall into reasonably defined categories, but each individual and every pet–owner interaction has aspects peculiar to itself. The classification of some of the more common pet–owner relationship types that follows is therefore a guide only.

The unwilling owner

Relatives or neighbors sometimes find that they are morally obliged to adopt an animal if the owner has died or has been admitted to institutional care. Despite feeling indifferent or even antagonistic towards the animal, they feel obliged to bear what they consider to be a burden. Initial resentment may even become guilt that they cannot relate more closely to the animal or provide for it as its owner did.

The "Afraid of the veterinarian" owner

Many people may have had their only contact with a veterinarian when their pet was euthanized. For the elderly this may relate to a time when in fact veterinary medicine was less advanced and euthanasia was the most common treatment for serious disease.

A deep-rooted fear of the veterinarian may be present such that bringing an animal for consultation is tantamount to inviting euthanasia. Equally, if euthanasia of a child's pet is handled clumsily, that child may have a long-lasting fear of veterinarians. Such attitudes may make establishing a veterinarian–client relationship difficult.

The multiple pet owner

Some individuals collect pet animals in what seems to be a compulsive fashion. In the more extreme cases, the numbers of animals in the household may be well beyond the number that can be kept in acceptable conditions on the owner's resources, threatening the health of both the animals and the owner. It is not uncommon for such persons to neglect their own welfare and diet in an attempt to fulfill the hopeless task of providing for the animals in their care. At the same time, the diet and environment of the animals may be deficient due either to lack of resources or to the owner's incapacity to understand or cope with the problems involved.

In one case of the author's experience, cats were regularly exsanguinated by a massive flea burden, to the great distress of the owner who was nonetheless unwilling or unable to apply even the simplest flea control measures.

Of course, not all individuals who own large numbers of animals should be regarded as suffering from some sociopsychological problem, and many will be found to be engaged in extensive voluntary work for disadvantaged humans as well as for homeless animals. However, too much animal orientation may lead to neglect of interactions with other people, and while this tendency may occur in individuals who own only one or two pets, it is probably found more often in multiple-pet owners. Such individuals may view the veterinarian as a fellow "animal lover" in a cruel world, and he may be much more trusted by them than their own doctors or social workers are. The veterinarian thus has a social obligation to such individuals and should, without betraying his professional trust, cooperate when necessary with social or psychological workers to aid the management of these owners.

The neurotic owner (the neurotic pet)

From the viewpoint of the medical practitioner, Mary Yoxall (Yoxall and Yoxall, 1979) has observed that well-meaning friends or relatives may give a pet animal to a slightly disturbed individual in order to "calm him down." It is sadly unlikely that a wholly satisfactory relationship will be established in such a case. The owner's problems may be exacerbated by the additional responsibility of pet ownership, while the animal is likely to develop behavioral problems. The resolution of such problems is likely to call for cooperation between those involved in the management of the owner and the veterinarian if the situation is not to be worsened by any attempt to remove the animal.

The hyperanxious owner

Hyperanxiety of the owners may stem from problems outside the pet-owner bond; but the same signs will be seen if it is directly related to the characteristics of the bond, and this will be discussed first.

It may be difficult to recognize in the examination room that an owner's stubborn or uncooperative attitude may be due to anxiety rather than a wilful intent to ignore the veterinarian's recommendations. A certain level of anxiety makes decision making extremely difficult; this level is different for each person, so the pet owner may require an unconscionable time to reach a decision that would appear to the veterinarian to be quite straightforward.

There are many reasons why owners become excessively anxious about their pet's welfare. They are often fully aware of their anxiety, and one frequently stated reason is that the animal has become a child substitute. It has been suggested (Yoxall and Yoxall, 1979) that young married couples will frequently acquire a pet animal once they have set up a home but before they feel ready to have children. The shared responsibility for the care and welfare of an animal might be a means of expressing a sense of family. This suggestion is supported by the observation of Katcher (1979) who noted that pet ownership was uncommon among unattached students at the University of Pennsylvania but that the incidence rose markedly among students who had settled relationships and were living together. In most cases, children become more of a focus for this "family feeling." Where the hoped for birth does not occur or where for some reason the parent–child bond is inadequate, the companion animal may easily become the recipient of the emotions that would normally be lavished upon a child. Yoxall and Yoxall (1979) have described another situation in which a similar role may be borne by a companion animal. This occurs with working women, especially when unmarried. Such individuals may devote all their energies to their work during their working lives but look forward to and plan for their retirement, intending then to care for an animal or animals onto which their frustrated maternalism may be poured. The degree of preplanning that is involved is likely to add to the importance of the human–companion animal relationship for such individuals. A similar phenomenon may be seen in women who have a strong maternal drive but their children have grown up and left home.

It should be remembered that the notion of pets as child substitutes has received popular acceptance due to the writings of people such as Konrad Lorenz (1964). Owners may feel better able to offer this explanation for their anxiety rather than dig deeper into their own particular situation.

As stated previously, hyperanxious owners frequently appreciate that they have a very high degree of concern for their animals and are often willing to discuss their motives, providing the veterinarian with an opportunity to lay to rest, at least temporarily, some deep-seated fears or anxieties. Occasionally, hyperanxious owners use their pets to obtain advice and information about their own health matters, the "ticket for admission" phenomenon discussed later. However, hyperanxiety may reflect experiences with previous pets, personal identification with the animals, or fear of a fatal condition. If the veterinarian can unearth the cause of such hyperanxiety, he should be able to reassure the owner that all is well and in so doing establish a trust that will be of value in subsequent dealings with the same client.

It is perhaps with the more straightforward cases of hyperanxiety that the practising veterinarian is able to do most to contribute toward the mental and emotional welfare of his clients. Ideally, one should aim for a history-taking method for new cases that will allow one to assess the relationship between the pet and the owner in the course of one's routine medical assessment of the patient. For the general practitioner it is not feasible to conduct a detailed investigation into each new case that one sees, but two or three "screening" questions can rapidly indicate which cases may deserve further attention. Leading questions can be used that elicit both physical, clinical information as well as psychological information in terms of the ways in which the replies are offered. Questions about feeding, exercise, sleeping accommodations, what the animal does during the day, what the animal does during holidays, and so on will reveal the degree of involvement of the owner with the animal and add to the general picture that one is able to form. The important point, especially with the hyperanxious client, is to keep the client talking, allowing as much opportunity as possible for expression of the problems underlying the fears and worries. This may take considerable time, but if the client's questions are unsympathetically handled or the flow of sometimes irrelevant details cut off, there is a good chance that the cause of the anxiety will never be revealed, and the veterinarian may have forfeited a good client relationship. Owner anxiety often leads to requests for surgery, rapidly followed by cancellation, requests for unnecessary diagnostic work or second opinions, noncompliance with instructions for medication, and a host of other problems that are likely to interfere with any potential benefits of treatment for the animal's physical condition. It must be stressed, then, that from the outset a favorable relationship with the owner must be established. In most cases, once one has secured a position of trust in the estimation of the hyperanxious owner, the owner's compliance with one's instructions is absolute, the anxiety being transferred to the risks consequent upon noncompliance.

If a satisfactory relationship is to be established, considerable time needs to be spent with any owner where the bond with his pet is abnormal. Some useful practical points that are worthy of consideration in any practice emerged from a meeting held in 1979 (Group for the Study of the Human–Companion Animal Bond, 1979). One of the most common handicaps was how to find time to deal adequately with cases of this nature in a busy small animal practice. Most of those who have attempted to undertake work of this sort within small animal practice agree that it is impossible to do very much during a short routine examination. Clients must be requested to return for a longer consultation. Most readily accede to this suggestion and are often more than a little pleased that one is prepared to set aside additional time and make extra efforts on their behalf. Very few cases won't wait for periods of up to twenty-four hours. Although such consultations will have to be fitted in after normal surgery hours, the clinician can view this additional burden upon his time in the same light as having to carry out an emergency surgical procedure outside his normal working hours, and the advantages of this sort of arrangement are several. Additional family members can be brought to the consultation who may cast vital light on the problem. Both clinician and client have a few hours to prepare themselves, and the clinician has the opportunity to seek advice from other sources. The client will be aware that this is an additional service for which a reasonable fee will be payable. It is worth noting that most clients are quite willing to pay for one's services in this context, and such activity should not be regarded as loss-making social service or goodwill exercise. Clients are probably more inclined to follow advice for which they have paid! In return, the veterinarian should normally be prepared to listen rather than to offer advice on anything other than the normal care or medical management of the animal concerned since, as with many purely human problems, "talking them out" can be a substantial part of resolution.

In a large practice with several veterinarians, there will often be one individual who is particularly able or willing to deal with such cases, and the suggestion that the client return for a "special" consultation with him or her will again show that the practice is taking special care over the case. Lay staff also have a potentially important role in this respect. In some practices in the United States, the veterinary receptionist is a person trained in human behavior observation who records routine details on the case notes, such as age, breed, sex of the patient, client's name and address, and also details such as number of individuals in the family (with some indication as to their relative ages) and the name of the family doctor "in case any zoonotic problem should occur." The client's interview with the receptionist might be brief or prolonged, the intention being to allow the client an opportunity to express his or her feelings about the animal in a relaxed and informal atmosphere with a sympathetic person who is not distanced from the client by factors such as professional standing, socioeconomic status, or the capability of prescribing euthanasia – factors that are likely to impede communication between the veterinarian and the client. The receptionist would then be able to provide the veterinarian with a synopsis of the potential pet–owner relationship problems involved in any case prior to his seeing the owner. In some cases, the receptionist or lay staff present may be able to resolve problems, particularly with hyperanxious clients during these preliminary discussions.

INDIRECT INTERACTION PROBLEMS

These are problems involving the pet–owner bond but arising externally to it.

Hyperanxiety

Hyperanxiety has previously been discussed in relation to its occurrence resulting directly from the pet–owner bond. Owners can, however, appear in such a state over their pet whereas in fact their anxiety is rooted elsewhere. It is naturally important for the veterinarian to understand this.

An animal may be presented for repeated examination for an ill-defined condition, and while this may represent the client's genuine fear that something is wrong with the animal it may equally represent their fear that they themselves have some disease.

Anxiety may be a manifestation of various other psychological phenomena. A proportion of owners regard their pets as "valued objects" or status symbols (Wilbur, 1976).

They may react in an extreme fashion to any threat to the animal as to any other valued object held in similar esteem. Certain breeds with a higher social image such as Burmese or Abyssinian cats or Afghans and salukis are likely to be very well represented in any classification of this sort. Some owners are anxious to show that they are sensitive and "caring" people by making highly vocal and emotional expressions of concern regarding their animals, and their response to pet ill-health, real or imagined, is similarly hyperanxious.

A more serious problem arises when the pet is valued as a link with a deceased relative. In an extreme case encountered by the author, a crisis for an elderly man arose when his cat became terminally ill. It transpired that the cat had originally been the pet of his wife, and he had failed to reconcile himself to her death while the cat was alive. His wife had died twenty years previously.

Because so many different causes produce the same signs, it is important to evaluate cases of hyperanxiety with care. Those that are related directly to the pet–owner bond may be more-or-less resolved by discussion with and simple reassurance from the veterinarian, while those that have deeper roots may require recognition and consultation with the appropriate doctors, psychiatrists, or social workers. A rather small proportion of owners is likely to be hyperanxious about everything, and the problem here is that if their anxiety is deflected from the animal, they are likely to reappear with anxieties related to payment of their bill and sundry other matters!

The "ticket for admission" phenomenon

This phenomenon was described by McCulloch (1976). In such cases, the owner has a need for professional help to deal with a real or imagined medical, psychological, or social problem but feels unable to approach the family doctor or other member of the health services directly. They then present their animal to the veterinarian with some trivial or fabricated complaint, perhaps because the veterinarian is a professional person who can provide direct advice or support on medical matters yet presents no direct threat or can make a difficult decision for the client with the authority of an informed professional bystander. Hyperanxiety may or may not be present, but it may be extremely difficult to identify these cases if the owner is at all reticent about expressing his real reason for seeking veterinary attention. This underlines the importance of getting the client to relax and talk. Commonly an owner may consult with the veterinarian if a close relative or friend has died perhaps away from home, and the owner has not had the circumstances adequately explained to them. They are thus seeking a "medical" discussion with an informed but disinterested third party to clarify certain points that worry them.

Pet animals in marriage

Marriage guidance counsellors report that pets are often involved in marital warfare because either the animal is valued by one partner and is thus abused by the other or it is used as a bargaining counter. Pet ownership may contribute to marital problems if one spouse feels that the other devotes too much affection to their animals or if one regards pet ownership as essential and a normal part of family life while the other, having been brought up in a family in which pets were regarded as at best a nuisance and at worst dirty and unhygienic, regards pets in a totally different light.

Pet abuse and child abuse

Within the context of the family as a whole, a social worker particularly involved in the problem of child abuse has observed a worrying number of cases in which families known to him because of a child abuse problem have presented their pet animals at local veterinary clinics with injuries of unexplained origin, often attributed to road traffic accidents. Such cases apparently tend to be taken to different local veterinary clinics after each incident so that no one veterinarian becomes aware of a history of recurrent injury (Hutton, 1978).

On the other hand, injuries to pets may be inflicted by children either unintentionally or intentionally (Ryder, 1973; Wax and Haddox, 1974; Justice, Justice, and Kraft, 1974). The practising veterinarian should be alert to the possibility of deliberate injury when attending to injured animals, especially if there is a history of recurrent injury. Pet abuse may be no more easy to discern than child abuse.

The pet–child bond

The specific relationships that develop between children and companion animals have been discussed elsewhere (McDonald, 1979). One of the problems in dealing with children's pets is that the animal is usually brought to the surgery by an adult, and information provided and attitudes registered may not reflect the specialness or closeness of the relationship between the child and the animal. Levinson (1972) has suggested that a companion animal may be an important source of support for children in families in which the relationship between the child and his parents is not fulfilling for the child. In this situation, as in so many others, an important insight into the relationship between the animal and the family as a whole is obtained by visiting the animal at home. A willingness to undertake home visits must be regarded as a prerequisite to making a useful contribution to pet ownership problems. The atmosphere of the veterinary clinic is a barrier to relaxed discussion of problems, while important clues as to the nature of the relationship between an animal and its owner can be obtained simply by entering the home.

Other psychosocial problems

Some owners may display inappropriate aggression or anger over trivial points arising from the medical management of their pets. The causes for such reactions are numerous. The owner may feel guilty that some real or imagined delay in presenting the case was the cause of an unsatisfactory outcome. The cause of the reaction may be conflicts experienced in work or in the family that are displaced in this way. Outbursts of anger may be a reflection of a certain personality style that may be most manifest when dealing with those in authority, including the veterinarian (McCulloch 1976), or may result from hardships at work, in the home, or in the family finances.

CONCLUSION

With increasing experience, the practising veterinarian learns how to handle most of these situations. He uses his trained abilities to observe and augments them with plain common sense. Until very recently, the veterinary schools have not helped him. In the future as more veterinary schools incorporate an understanding of the human–companion animal bond into their curricula, the graduating veterinarian will have an understanding of the pet–owner interaction to a more sophisticated degree than today. In addition, a return to a more relaxed consultation style and the use of lay staff trained in human behavior will be important aids to adequate handling of problems of interaction.

REFERENCES

Argyle, M.: *The Psychology of Interpersonal Behaviour*. London, Penguin, 1967.

Group for the Study of the Human–Companion Animal Bond. *Proceedings of the First Meeting, Held at the University of Dundee*. March, 1979.

Hutton, personal communication, 1978.

Justice, B., Justice, R., and Kraft, I. A.: Early warning signs of violence: is a triad enough? *Amer J Psychiat, 131*: 457–9, 1974.

Katcher, A.: Social support and health: effects of pet ownership. Group for the Study of the Human–Companion Animal Bond, *Proceedings no. 1*, 1979.

Levinson, B. M.: *Pets and Human Development*, Springfield, Thomas, 1972.

Lorenz, K. Z.: *Man Meets Dog*. London, Penguin, 1964.

McCulloch, M. J.: Contribution of veterinarians to mental health. In *A Description of the Responsibilities of Veterinarians as They Relate Directly to Human Health*. Washington, D.C., U.S. Government, Department of Health, Education and Welfare, Bureau of Health Manpower. Contract No. 231-76-0202, 1976.

McDonald, A. J.: Review: Children and Companion animals. *Child: Care, Health & Develop, 5*: 347–58, 1979.

Ryder, R. D.: Pets in man's search for sanity. *J Small Anim Pract, 14*: 657–68, 1973.

Wax, D. E. and Haddox, V. G.: Enuresis, firesetting and animal cruelty: a useful danger signal in predicting vulnerability of adolescent males to assaultive behaviour. *Child Psychiat Hum Develop, 4*: 151–6, 1974.

Wilbur, R. H.: *Pet Ownership and Animal Control: Social and Psychological Attitudes*. Report to the National Conference on Dog and Cat Control, Denver, Colorado, 4th. Feb., 1976.

Yoxall, A. T. and Yoxall, D. M.: The Multi-disciplinary approach to problems of the pet-owner bond. Group for the Study of the Human–Companion Animal Bond, *Proceedings no. 1*, 1979.

CHAPTER 18

ATTACHMENT–EUTHANASIA–GRIEVING

BRUCE FOGLE

> *When God created the world, He evidently
> did not foresee the future bond of friend-
> ship between man and the dog.*
>
> Konrad Lorenz

E very day, in almost every veterinary practice in the world,
veterinarians make decisions to terminate lives. Of all the
areas of medicine, this is one that is unique to veterinary medicine.

But how does one make such a decision? How does one develop
the technical understanding of a case, the understanding of the
relationship between the animal and its owners, the understanding of that animal's innate will to live? How does the veterinarian
ever know enough to make a verdict of doom? And how does he
carry it out?

Euthanasia in veterinary medicine is an everyday practice. In a
survey of seven western American veterinary practices, 2.1 percent
of all client visits to the veterinary clinics resulted in completed
euthanasias (*see* Chapter 6). The same writer has previously estimated that there are 125 million "animal visits" to small animal
veterinarians in the United States yearly (McCulloch, 1979). If
that is so, then each year veterinarians in the United States euthanize
2.5 million dogs and cats. And that only accounts for euthanasias
carried out by veterinarians. To that number must be added the
millions of pets, not feral dogs and cats but pets, that are euthanized
by local humane societies.

Interrelations Between People and Pets

ATTACHMENT

*Near this spot are deposited the remains
of one who possessed beauty without vanity,
strength without insolence courage without ferocity and all the virtues of man,
without his vices. This praise which would
be unmeaning flattery if inscribed over
human ashes is but a just tribute to the
memory of Boatswain, a dog.*

Lord Byron

Any discussion of euthanasia of pets must be based on an understanding of the attachment that develops between owners and pets. Attachment is of crucial importance in all social animal species. Human development, however, inevitably brings with it a distrust of permanent unwavering attachments. The unmitigated attachment that a baby has for its mother must finally yield to permit the independence necessary for intellectual development.

But we undoubtedly remember the more pure form of attachment. Because of this, one writer suggests that "the exchange of acceptance and affection between us and pets is less complicated than human exchange of similar need and satisfaction (Rynearson, 1978). Another psychiatrist says that "the animal does not judge, but offers a feeling of intense loyalty. . . . it is not demanding. . . . it does not expose its master to the ugly strain of constant criticism. It provides its owner with a chance to feel important" (Siegal, 1964). Yet another clinical psychiatrist states, "Animals . . . bolster the pet owner's morale and remind him that he is, in fact, a special and unique individual" (Keddie, 1979).

We don't need psychiatrists to tell us this. In 1930 in a perceptive little book called *Lets Talk of Dogs*, Rowland Johns put these words into a Hyde Park orator's mouth as the orator spied a man in the crowd holding a dog.

You come home at night after a row with your foreman, you're wet through because you've lost your return tram ticket and the gee-gee you backed in the two-thirty yesterday wasn't a winner. You're feeling like a human dish-cloth, your morale is in shreds and there isn't a soul in the world who would care if you fell down a man-hole and got into the river via the main sewer. The wife is fed up with you. The children wonder however you came to be their dad, nobody wants you, nobody cares.

And then your little dog—I expect his name is Jack—sits up and watches you as you hog your mutton chop, and comes closer to lick your hand. He puts his chin—I suppose dogs have chins—on your knees, and looks at you with devotion in his eyes. He tells you that you are the finest chap that he knows, that everyone else misjudges you. The missus used to say that sort of thing years ago, when she didn't know you so well. And you tell yourself that Jack is your real pal. He never tips you a loser, never wants your wages, never grumbles when you want company on a walk in the park. Your dog gives you new courage, he helps you change from a half-dead flat-fish into a new man. And you put yourself on a pedestal as a friend of his.

Reasons for attachment

There are several factors today that are making this bond of more importance to the veterinarian. The classic family unit of two parents with children may no longer be the dominant unit in western society. And with its demise goes the security that it brings. In a survey in a Toronto veterinary practice in a middle-class neighbourhood, only 31 percent of the clients were from the "typical" family of two parents with children. Single parent families accounted for 4 percent of the clients seen; 23 percent were unmarried pet owners living alone; 24 percent were older couples whose children had left home (empty nesters); and 18 percent were childless couples (McCutcheon, 1979).

Compounding this change in family structure is the decreasing influence of the other totems of our previous life-style. The extended family of grandparents, parents, children and relatives living in proximity to one another is almost a thing of the past. Religion is no longer a significant force in the community. These things previously provided people with companionship, security and, above all, human contact. As they have been lost, pet animals have quietly slipped into the void.

But there are other factors that influence the attachment between people and their pets. It appears that survival time of individual dogs and cats is lengthening (Schneider, 1979). And the longer one has a dog or cat as a pet, the greater the attachment can become. When a client writes to me after the death of a pet, the age of the pet is frequently mentioned. "Fifteen years is a chunk of one's life and more of the children's." "Parting after nearly twelve years is bitter." "We miss Luke very much. Fifteen years ago when he came into our home Nicola was in Nursery School. Now she is married."

The emotional strain can be so great that 15 percent of pet owners when asked whether they would get another pet after the present pet died said they would not because of the psychological trauma that followed the present pet's death (Wilbur, 1976).

Compounding all of these facts is the simple idea that genetically we are programmed to live in a symbiotic fashion with nature. Our genetic evolution took millions of years, and we developed to live in close relationship to the rest of creation (Wilson, 1978). It is only in the last 20,000 years that we have hived ourselves off from our roots. Chief Sealth of the Duwamish Tribe in the state of Washington was perhaps saying more than he knew when he wrote to the President of the United States in 1855. "What is man without beasts? If all the beasts are gone man would die from great loneliness of spirit, for whatever happens to the beasts also happens to man. All things are connected. Whatever befalls the earth, befalls the sons of the earth." Was Chief Sealth the first sociobiologist?

EUTHANASIA

> *Few people have the opportunities in their lives to genuinely help as many people as the small animal practitioner can help in a day. By help, I mean help with the emotional decisions that tug at the very roots of the part of us that sets us apart from the animals we treat.*
>
> *Hopkins, 1978*

Veterinarians have ample literature to call on, to ascertain which are the best methods of euthanasia. Indeed, in the last decade, the veterinary associations and humane societies in the United States, Canada, and the United Kingdom have all published reports and guidelines on euthanasia (AVMA, 1978; CVMA, 1978; RSPCA, 1979). Others have published equally detailed reports (Urquhart, 1976; Carding, 1977; Rowsell, 1979), but none have dealt even peripherally with the equally important question of how to make the decision. And it is a question that until recently certainly hadn't entered the curricula of the veterinary schools.

The word *euthanasia* is derived from Greek and means death with peace. Because it is almost a taboo subject in western society, its discussion provokes strong opinions, opinionated opinions, among veterinarians as well as pet owners. Many pet owners are far more distressed at the thought of euthanasia, an unnatural death, that at the thought of prolonging life with assisted ventilation and kidney dialysis, an unnatural life. "I bawled my eyes out more than I thought possible but I'm glad she died naturally and didn't have to be put down. That would have seemed much worse." But it is only natural to dread euthanasia. It is man-made. It is permanent. There is no recourse.

Legally, companion animals are, in most western countries, defined as property or chattels and as such, the owners can do with them as they please as long as other laws are not broken. (This may be changing. A new law in California prevents malicious and unnecessary destruction of animals. This law was tested in June, 1980 when a group called Attorneys for Animals' Rights successfully used the law to prevent the destruction of an animal whose owner had stipulated in her will that after her death the animal be destroyed).

Reasons for euthanasia vary, but generally the veterinarian is servant to the owner's wishes for his services, and in the case of euthanasia, the owner has the legal right to ask a veterinarian to kill his pet. The veterinarian has the right to refuse.

Hopkins (1978) describes the complexity of this situation by assigning six different sets of owners the task of deciding what to do with an old pet cat with lymphosarcoma. With treatment it may live two to twelve months. Without treatment it will die within two months. Hopkins outlines the three choices (euthanasia, treatment, or delay of euthanasia) that these couples have and then explains how one choice would apply to the young childless couple and another to the single parent family where the parent works. His actual choices are not important because they cannot be hard and fast. What is important is that under the same medical circumstances he might do three different things.

There are no rules as to how this decision can be made, but there can be guidelines. Hershhorn, in his book *Active Years for Your Aging Dog*, gives six criteria to help owners make a decision (Hershhorn, 1978).

1. Is the condition prolonged, recurring, or getting worse?

2. Is the condition no longer responsive to therapy?

3. Is your dog in pain or otherwise physically suffering?

4. Is it no longer possible to alleviate that pain or suffering?

5. If your dog should recover, is it likely to be chronically ill, an invalid, or unable to care for itself as a healthy dog can?

6. If your dog recovers, is he likely to no longer be able to enjoy life or will he have severe personality changes?

If the answer to all of these questions is "yes," then euthanasia should be undertaken. If the answer to three or four of them is "no," then the pet should be permitted to die naturally, but the following questions should first be answered by the owner.

1. Can you provide the necessary care?

2. Will such care interfere with your own life as to create serious problems for you or your family?

3. Will the cost involved become unbearably expensive?

"There is such a disparity between living and being kept alive", says Hershhorn. It is the "quality of life" that matters.

Reasons for Euthanasia

There are many reasons why I am called upon to perform euthanasia. They include—

1. Overwhelming physical injury.

2. Irreversible disease that has progressed to the point where distress or discomfort cannot be controlled.

3. Old age wear and tear affecting the "quality of life."

4. Physical injury, disease, or wear and tear resulting in permanent loss of control of body functions.

5. Aggressiveness with danger to the owners and others.

6. Pet carries disease dangerous to man.

7. Pet is a nuisance through its behavior to its owners or neighbors.

8. Owners are allergic to pet.

9. Owners cannot financially support pet.

10. Owners change home to new residence that cannot or will not accept pets.

11. Owners want a change; want to try something new.

12. Owners "can't bear to see an animal suffer."

In some of these categories, there are alternatives to euthanasia. In most there are not, and when the time comes what does one actually do? In passive euthanasia in man, although the doctors make the decision to pull the plug, the poor nurse is usually left with the task. In active euthanasia in animals, it is wholly the veterinarian's responsibility.

The Act of Euthanasia

A basic need is for a firm understanding with the owners. Communication is usually best within one's peer group, and as veterinarians are firmly middle-class they communicate best with middle-class owners. Working-class owners, for many reasons, frequently expect the veterinarian to make the decision, whereas middle-class owners are more likely to identify when the time for euthanasia has come.

In my practice, once a decision has been made, I ask the owners whether they would like to stay or leave. If there is hesitancy, I make the decision for them. I do not ask that a "release form" be signed. The British are not litigation conscious, and professional liability insurance covers this field. To sign a release form induces an unnecessary feeling of guilt in the owner.

If the owner has not stayed in the room, he or she is invited to return to see that the animal is dead. It perhaps helps them to appreciate the reality of the event, and because the animal looks "peaceful" it may enhance a feeling of doing the right thing. Children may need to see an animal not breathing to understand death. (Adults frequently bring their children's ill pets in for euthanasia to shield the children from the event. I think this is wrong. Because of their lack of familiarity with death children may not understand that the pet has died. They may think rather that it has simply been taken away from them.)

The owners are encouraged to stay, talk to the nurses, and perhaps have a cup of coffee. The nurses, although they may have assisted with the euthanasia, have not been involved in the decision making. Very often the owners are more at ease with them at that time than with the veterinarian (or with themselves). Traffic is chaotic in Central London. We try to make sure that owners leave with unimpaired vision.

After Euthanasia

But what happens when they get home? And should it concern the veterinarian?

Yoxall reminds us that very often owners cope very well at the veterinary clinic, but it is "when they return home to face the empty basket, the empty food bowl, or the absence of the pet in the home that full realization of the loss makes itself felt" (Yoxall, 1979).

The following day I write a note to the owners reassuring them that the decision they made was correct, humane and, most importantly, unselfish.

The veterinarian must prepare both himself and his clients for the prospect of euthanasia of pets. Frank discussion of euthanasia (and disposal of the body) should be held with the owners long before a decision has to be made. The veterinarian should develop a backup of "sidewalk consultants" upon whom he can call for advice. In certain cases, he should have the means of contacting the owner's neighbor or doctor for support after euthanasia.

Veterinarians have a long experience in dealing with euthanasia, in dealing with death in a way in which doctors do not. Death is dealt with more frankly and directly by veterinarians. It is a day-to-day routine, and it the most difficult part of practice.

GRIEVING

> *The vet looked at me, then I put my arm around my wife and led her out to the car. We talked on the way home, but our voices sounded strange to each other, and we didn't really say anything. I think we must have felt as parents do when they have lost a child. . . . I am not a man who cries easily, but picking up her bone and her rubber ball, her food and water dishes, I cried as I haven't cried since childhood. I'm glad my wife wasn't there to see me cry that way, because I would have had to stop then, and the hurt within me would have been worse.*
>
> *Joseph, 1957*

What is normal grieving? What is pathological grieving? And what should the veterinarian know about these things?

Some degree of mourning after bereavement is necessary and normal. A pet owner's reaction to the death of his pet resembles in many respects the normal grief response following the death of a loved one (Keddie, 1977). And the normal grief response to a loved one's death is well documented.

Normal Grieving

The end of a person's life is final for the deceased but not for the survivors. For friends and relatives, acceptance and adjustment to the loss is a protracted and painful process in which the lost object is given up gradually and often with a struggle (Bird, 1973). The first reaction may be conscious or unconscious denial of the death. Memories of the deceased may seem sharpened. "In the weeks immediately following Bully's death, I really began to

Interrelations Between People and Pets

understand what it is that makes naive people believe in the ghosts of the dead. The constant sound throughout 17 years of the dog trotting at my heels had left such a lasting impression on my brain that for weeks afterwards as if with my own ears, I heard him pattering after me." (Lorenz, 1957). There is a sadness, a sense of loss, of being lost, and of hopelessness. Concurrently there is a lessened interest in one's surroundings. "Gradually however, bit by bit, memory by memory, death really occurs. The object is given up and only as this happens can healing take place and new objects be found to replace the old" (Bird, 1973).

Many people seek religious support when grieving, and this applies to grieving at pet loss as well. The Royal Society for the Prevention of Cruelty to Animals produces a booklet entitled *Service of Prayers for Animals.* Dogs are occasionally buried with full religious ceremony. When a client, a vicar, lost his pet recently, he wrote to me, "Many years ago when an adored dog died, a great friend, a Bishop, said to me, 'You must always remember that as far as the bible is concerned God only threw the humans out of Paradise.' "

The grief is real. It must be respected even though others may not understand. And it falls upon the veterinarian to provide that understanding when he can.

Pathological Grieving

If, however, the emotional investment in the pet is intense, the death of that pet can create complicated grief, in part because of denial of the death but also because of the shame in admitting such intense nonhuman attachment. McCulloch has recounted a case of a middle-class teenager, an academic high achiever whose academic performance deteriorated one month after the accidental death of his pet Alsatian. In ten weeks after the death, the problem was severe enough to warrant referral for psychiatric evaluation. During this it was discovered that his personality and perform- ance changes were related to his feeling of shame at grieving so deeply over the loss of "only a dog" (McCulloch, 1979). Keddie feels that pathological mourning is liable to occur in circum- stances where there has been "undue reliance on the pet, usually to the detriment of meaningful human relationships" (Keddie, 1979). He says that the greater the degree to which the pet has been placed in a "quasi-human" role during its lifetime, "the greater will be the risk that the owner will suffer a possibly emotionally damaging type of reaction after the pet has been euthanized or accidently killed" (Keddie, 1977).

Keddie feels that pathological mourning is most common in relation to deaths of small pets (Keddie, 1979), although that is not my experience. I have seen what appears to be excessive mourning in owners of large and small dogs equally. The age of the animal seems directly related to the degree of grieving, however. I have never seen more sorrow than that shown at the death of a thirty-five-year-old tortoise.

Grieving at the death of a pet may appear exaggerated when it is related to a previous death of a friend or relative and that grieving has been suppressed. In circumstances such as these, the owner sometimes sees the pet as an animate link with the person who died. The pet's death triggers off grieving for the deceased and for the pet—double mourning (Yoxall, 1979). A further complication that the veterinarian may encounter involves the elderly. Some pet owners, especially the ill and the elderly, look on their pets as part of themselves. A decision to euthanize is taken personally. A parallel is drawn with their own situation in life.

When one understands the stages of grieving, one then questions whether encouraging owners to obtain new pets after the death of the old is right or wrong. With strong attachments, some owners will need time in which to work through the grief. "Somehow we feel she will never really leave us. She seemed to have always been here and always will be."

Because of the biological attachment we feel for our pets, because with time they become personalities and not simply hairy quadrupeds, because of the time we spend playing with them and caring for them, we grieve over their loss as we would over a friend's. "It is always hard to lose a good friend and Moffet certainly was that." "He was a little person to Maggie and me and we miss him very much." "She was never just a family pet as some animals chose to be. She was my friend." "We are still not accustomed to the absence of that very intelligent and undemanding creature who had been a faithful and trusting friend for over thirteen years." These are not letters from excessive mourners. They are from typical, solid, well adjusted pet owners. By having these creatures in our homes for so many years, they become part of our lives. The veterinarian needs to know this, for whether he likes it or not he has been drawn inextricably into the social care of his clients. Albert Kushlick describes two types of social care (Walster, 1979). He distinguishes between "hit and run" people and "continuing care" people. The nature of the small animal veterinarian's work, the prevention of disease, the treatment of illness and trauma, the bringing into the world and the taking away of many successive generations of patients, and the care of the owners places him or her in a position once held by the local doctor and priest. Now in its own unique way it is held only by the veterinarian. May he and she continue to be classified as "continuing care."

REFERENCES

AVMA: Report of the AVMA panel on euthanasia. *J Amer Vet Med Ass'n, 173*: 59–72, 1978.

Bird, Brian: *Talking with Patients*, 2nd ed. Lippincott, Philadelphia, 1973, p. 74.

Carding, T.: Euthanasia of dogs and cats. *Animal Regulation Studies, 1*: 5–21,1 977.

CVMA: The euthanasia of dogs and cats: statement by the humane practices committee of the Canadian Veterinary Medical Association. *Can Vet J, 19*: 164–68, 1978.

Hershhorn, Bernard: *Active Years For Your Aging Dog*. Hawthorn Books Inc., New York, 209–22, 1978.

Hopkins, A. F.: In McCullough, L. B. (Ed.): *Implications of History and Ethics to Medicine — Veterinary and Human*. Ethical Implications in Issues and Decisions in Companion Animal Medicine. Texas A&M University, College Station, Texas, 107–14, 1978.

Johns, Rowland: *Lets Talk of Dogs*. Methuen & Co., London, 94–98, 1930.

Joseph, Richard: *A Letter to the Man Who Killed My Dog*. W. H. Allen, London, 1957.

Keddie, K. M. G.: Pathological mourning after the death of a domestic pet. *Brit J Psychiat, 131*: 21–25, 1977.

_____ : *Proceedings of Meeting of Group for the Study of Human-Companion Animal Bond*. Dundee, Scotland, March 23–25, 1979.

Lorenz, Konrad: *Man Meets Dog*. Methuen & Co., London, 193–99, 1977.

McCulloch, Michael: *Proceedings of Meeting of Group for the Study of Human-Companion Animal Bond*. Dundee, Scotland, March 23–25, 1979.

McCutcheon, Paul: Psychological aspects of the family relationship. *Proceedings of the Second Canadian Symposium on Pets and Society*, 3–6, 1979.

Rowsell, H.: Euthanasia, the final chapter. *Proceedings of the Second Canadian Symposium on Pets and Society*. 125–39, 1979.

RSPCA: Report of the ad hoc advisory committee on euthanasia. *Vet Rec*, 104–8, 171, 1979.

Rynearson, E. K.: *Humans and pets and attachment. Brit J Psychiat*, 133, 550–55, 1978.

Schneider, R.: Pet ownership: some factors and trends. *Proceedings of the Second Canadian Symposium on Pets and Society*. 142–44, 1979.

Siegal, A.: Reaching the severely withdrawn through pet therapy. *Am J Psychiat*, *118*: 1045–6, 1964.

Urquhart, R. G.: Euthanasia, a necessary evil. *Proceedings of the First Canadian Symposium on Pets and Society*. 81–84, 1976.

Walster, Dorothy: *Proceedings of Meeting of Group for the Study of Human-Companion Animal Bond*. Dundee, Scotland, March 23–25, 1979.

Wilbur, R. H.: Pets, pet ownership and animal control. Social and psychological attitudes. *Proceedings of the National Conference on Dog and Cat Control*. American Humane Association, Denver, Col., February 1976.

Wilson, Edward: *On Human Nature*. Harvard University Press, Cambridge, Mass., 1978.

Yoxall, Andrew and Yoxall, Dorothy: *Proceedings of Meeting of Group for the Study of Human-Companion Animal Bond*. Dundee, Scotland, March 23–25, 1979.

INDEX

A

Abandonment of dogs, 75
Abuse. (*See* Child abuse; Pet abuse)
Academic inflexibility, 248–49
Actions, Styles, and Symbols in Kinetic Family Drawing (Burns and Kaufman), 189
Active Years for Your Aging Dog (Hershhorn), 336
Activity cycles of companion animals, 16
Actualizing relationships, 33
Affection and sexuality, 44
Affiliation, 115
African hunting dog, 13
Aged. (*See* Elderly)
Agee, James, 75, 81
Age of dogs, and behavior problems, 310–11
Aggression, 300–303, 308. (*See also* Biting; Destructive behavior)
Agricultural era, 25–27
Alcoholics, pet therapy for, 135
Alice in Wonderland (Carroll), 243
Allergic reactions, 163
Alsatians, 319–20
Altruism, 272
American Association for the Prevention of Cruelty to Animals, 180. (*See also* Pet abuse)
American Humane Association, 76
American Humane Society, 136–38
American Psychological Association Division, 131
Anatomy of an Illness as Perceived by the Patient (Cousins), 115
Anderson, R. K., 253
Animal
    children's drawings of, 188–93
    husbandry, 27
    -nature connection, 30
    phobias, 178, 227–28, 303–304
    relationships, and human community health, 264 (*see also* Therapeutic value of pets)
    (*See also* Pets)
"Animal Behavior in Veterinary Medicine," (Fox), 251
Anthropocentrism of scientific psychology, 42
Anthropological biases, 45
Anxiety. (*See* Hyperanxiety)
Apartment living with pets, 37. (*See also* Urban)
Ashton School for Handicapped Children, 137
Asia, role of dogs in, 70
Attachment theory, 178, 205, 272–82, 332–34
Attention-getting behaviors, 305–306
Attitude, and pets, 89–98
Attorneys for Animals' Rights, 335
Audiogenic psychomotor seizure, 276–77
Australian aborigines, 8, 11
Autistics, pet therapy for, 135
Automobile problems of dogs, 303–304

B

"Babytalk," 53–54
Balson, W. S., 160
Beaver, B., 252
Beck, A., 71, 142, 209
Behavior
    and attitude, 89–90
    fact sheets, 222–23
    modification, 296
    problems, 35
        aggression, 300–303
        Behavior Problems in Dogs (Campbell), 186
        of cats, 292–93
        diagnosis and treatment of, 226–28
        in obedience, 305
        owner-caused, 222, 318–26
        physiological, 304–305
        prevention of, 283–93

relative frequency of, 306–311
results of treatment for, 311–12
subtle reinforcement of, 298
Belaer, D., 19
Benefits of pet therapy, 170. (*See also* Pet therapy)
Bernard, C., 42–43
Between Animal and Man (Fox), 31
Bierce, A., 77
Biological tradition, 42–44
Biting, 163, 178, 182–84
case histories, 274–79
laws regulating, 233
owner encouragement of, 319
(*See also* Aggression)
Blind people's guide dogs, 248
Blood pressure, and speech, 54–56
Bolwig, N., 14
Bowlby, J., 44, 84–85, 178
Breed of dogs, 18–19, 35
and behavior problems, 308–310
for pet therapy, 134
Brickel, C., 139, 140, 154
Brickel, G., 140
Bridges, H., 195
Brodey, R., 213
Broken Heart, the Medical Consequences of Lone-
liness, The (Lynch), 116
Burns, R., 188–89, 191–93
Bustad, L., 136, 142
Butler, R., 146
Byron, G., 332

C

Campbell, W., 186
Canine
Companions for Independence, 138
Therapy Program, 137
waste laws, 233–35. (*See also* Dogs)
Canis familiaris, 12
Cannon, W., 43
Captive breeding, 11
Cardiovascular disease, 47–52
Carroll, L., 243
Case histories
of canine elimination problems, 227
communication between owners and vet-
erinarians, 217–18

of owner's attachment to pets, 273–79
owner exacerbation of pet aggression,
229–30
pet fears of owner, 227–28
pet therapy
for medical illness, 103–104
in nursing homes, 163–68
in psychotherapy, 185–88
Castle Nursing Homes (Millersburg, Ohio),
156–68
Castration, 35, 37, 225, 306–308
Cats, 14–18
behavior problems of, 226, 292–93
origins of, 6, 9, 11
sex of owners of, 60–61
as symbols in children's drawings, 191–92
therapeutic functions of, 140, 154
U.S. population of, 151, 237
Center on the Interaction of Animals and
Society, 137, 209
Character traits, and illness, 113
"Charlie Brown," 176
Charlotte's Web (White), 243
Chew toys, 291
Children
abuse of, 152, 180–82, 328
attachment to, 280
bonds of, with pets, 328–29
cruelty to animals among, 178–79
development of, and pets, 175–77
drawings of, 188–93
fear of animals among, 178
pet education of, 243–46
pets as substitutes for, 78–79, 322–23
Children's Aid society, 183, 184
"Children and Animals Together for Seniors,"
138
Child Welfare Act of Ontario, 183–84
"Chronic Disabling Illness: A Holistic View,"
(Feldman), 121
Chronicles of Narnia (Lewis), 243
City design. (*See* Urban)
Classroom pets, 244–46
Clutton-Brock, J., 12
Coats of animals, 17–18
Cochrane, B., 183
College of Veterinary Medicine, Pullman
Washington, 136, 142
Coming of Age, The (de Beauvoir), 146

Commensalism, 9
Communication between owners and veterinarians, 215–19
Companion Animal Clinic, University of Pennsylvania, 209–220 passim, 221, 225–28
Conditioned-aversion technique, 306
Condoret, A., 41, 46, 57, 64
Cooper, E., 142
Coprophagia, 306
Corson, S., 41, 46, 57, 127, 140, 142, 185
Cousins, N., 115–16
Crossbreeds, 308
Cruelty. (*See* Pet abuse)

D

Darwin, C., 44–45, 150
Davis, S., 151
Deafness, 135
Death of a pet, 32–33, 76, 119, 176–78, 186, 205, 339–42. (*See also* Euthanasia)
Debarking, 133
de Beauvior, Simone, 146
Declawing, 35, 37
Degradation of the elderly, 146
DeHass, D., 156, 164
Dehumanization, 147
Delta Foundation, 142
Depression, pet therapy for, 105–119
Deprivation techniques, 287–88
Descartes, R., 42
Destructive behavior of dogs, 37, 273–76. (*See also* Aggression)
Devocalizing of pets, 35, 37
Dhole, 13
Dingoes, 11
Direct interaction problems between owners and pets, 318–26
Discrimination against pets, 35
Docility, 18–19
Dogs
    age of, and behavior, 310
    behavior problems of, 35, 186, 225–28, 283–91, 298, 300–312
    biting by. (*see* Biting)
    in children's drawings, 193
    and child surrogates, 78–79

coat of, 18
"contradictory valuations" of, in Smith, 74–75
crossbreeds of, 308
definitions of, 74, 77
diurnal habits of, 16
earliest domestication of, 5–11
in European households, 231
expressiveness of, 14
fidelity of, 80–81
housetraining of, 16 (*see also* Elimination)
human ambivalence toward, 73–77
idealization of, by owner, 77–78
intelligence of, 15
leashing of, 233
morphological changes resulting from the domestication of, 17–18
origins of, 5–6, 151
as parents to their owners, 79–81
playfulness of, 15
primary socialization of, 12–13
role of, in Asian society, 70
runs, 235–36
size of, 61
as transitional objects, 81–82
in urban environments, 36, 36, 231–32
U.S. population of, 151, 231
Dog Crisis, The (Nowell), 193
Domestication, 6–11, 17–19, 25–27
Dominance, 13, 283–88
Do We Need Dogs (Adell-Bath), 193
Downs, J., 8–10
Drawings, in child psychotherapy, 188–93
Drug addicts, therapeutic animals for, 135
Dusicyon australis, 12
Dutch sennenhund, 301

E

Education about companion animals, 252
    of the general public, 255–57, 295–97
    in high school, 246–47
    of owners, 238, 286
    of teachers, 245
    for undergraduate students, 247–50
    for veterinary professionals, 241, 250–55

"Educational Use of Pets with Handicapped Youth," (Jacobson), 140
Egypt, 6, 8, 27
Ein Mallaha (Eynan), 151
Elderly
    depression in, 117
    and hearing loss, 148–49
    isolation of, 146–47, 150
    selection of dog for use with, 135
    society's attitude towards, 146
    suicide among, 117
    and therapeutic value of pets, 30, 102
Elimination, 16, 233–35, 303
Emotions of pets, 38
Empathy of animals, 52–53, 56–57
"Empty nest syndrome," 147
Enlarged Devil's Dictionary (Bierce), 77
Environmental management and design for pets, 235–37
Environmental setting, and pet therapy, 128–29
Epidemiology of heart disease, 50–51
Epipaleolithic period, 151
Ethical problems of pet ownership, 35–37
Ethology, 44–46, 254–55
European
    explorers, 6
    pet populations, 231
Euthanasia, 102
    for behavior problem dogs, 311–12
    as the difference between human and veterinary medicine, 219
    guidelines, 336
    high school studies of, 247
    incidence of, 102, 331
    legal status of, 335
    owner reluctance about, 275–83 passim
    social etiquette of, 337–38.
    (*See also* Death of a Pet)
"Evaluation apprehension," 54–55
Evans, W., 75, 81
Expenditures for pets in the United States, 151
Exploitative relations with pets, 31
Expression of Emotions in Man and Animals (Darwin), 44, 150

F

Fabre, J. H., 125
Facial expressions of primates and canids, 14
Failures in pet ownership, 295
Falconry, 26
Fear. (*See* Phobias)
Feces. (*See* Canine waste laws; Elimination)
Feldman, D., 121
Felis sylvestris, 6
Fidelity of dogs, 76–77
Finch, S., 154
Fishbein, M., 90
Flicka, 244
Food surplus, and pet breeding, 10–11
4-H clubs, 247–48
Fox, Michael, 56, 152, 175, 251, 296
Freiberger, R., 142
Freud, S., 43, 46–47
Friberg, A., 137
Friedman, E., 140
Friedman, M., 117
Fuller, J. L., 79–80, 152
Future Farmers of America, 247

G

Gantt, W., 153
Gender. (*See* Sex)
Geriatric institutions. (*See* Nursing homes)
Golding, W., 179
Goodman, M., 52
Graham, T., 156
Grahame, K., 243
Greeting behaviors, 280–81
Grieving, 339–42. (*See also* Death of a Pet)
Grooming, and dominance training of dogs, 287
Group for the Study of the Human-Companion Animal Bond, 296
Handicapped pet assistance, 134–35, 138, 248
Harlow, H., 44, 153
Harry the Dirty Dog (Zion), 243
Hart, B., 296
Hayonim Terrace, 151
Health hazards and pet therapy, 143
Heart disease, 50–51, 102
Hearing-ear program, 134, 138

Helfer, R., 180
Herbivores, domestication of, 16
Hershhorn, B., 336
Hess, E., 150
Hilzheimer, M., 9
Holmes, T., 149
Hopkins, A., 334
Horseback riding as therapy, 138–39
Housetraining, 16. (*See also* Elimination)
"House-Tree-Person" Test, 188
Humane Society of Colorado, 136
Humanistic education, 241–42, 249
Human services personnel, 209
Humor, 115–16
Hunter-gatherers, 5–6, 25, 151
Hunting, 8, 11, 26
Hygiene, and pet therapy, 132
Hyperanxious owners, 322–27

I

Ice Age, 10
Idea of the University (Newman), 241
Idle play, 62–63
Inattention to the meaning of pets, 41–42
Independence of nonpet owners, 92
Indiana University, 140
Indirect interaction problems between owner
    and pet, 326–29
Indulgence of pets, 285–86
Infancy, 44, 84–85
Institutional settings, 128, 141
Intelligence of cats and dogs, 15
Intimate Behavior (Morris), 65
Intraspecific attachment, 279
Involvement, 241, 243–45
Isolation of elderly in society, 146–47, 150
Isolation anxiety of pets. (*See* Separation anx-
    iety)

J

Jacobson, J. J., 140
Japan, role of dogs in, 70
Jekyll and Hyde syndrome, 302
Jordan, J., 74–75
Judeo-Christian morality, 42

Jungle Book (Kipling), 243
Junior League of Boston, 137–38
Juveniles, selection of therapeutic dogs for,
    135

K

Kane, Robert, 147–48
Kane, Rosalie, 147–48
Katcher, A., 137, 204, 209–210, 212, 253, 322
Kaufman, S., 188–89, 191, 192, 193
Keddie, K., 177, 341
Kempe, C., 180
Kinetic-Family-Drawing, 189
Kipling, R., 243
Klapka, Jerome, 156
Kushlick, A., 342

L

Laissez-faire philosophy of pet ownership,
    34
Lassie, 244
Leashing of dogs, 233
Lee, D., 139, 160–61
Legislation about pets, 141, 232–35
Let Us Now Praise Famous Men (Agee and
    Evans), 75
Levinson, B., 41, 46, 57, 140, 142, 152, 154,
    175, 182, 185, 195, 198–200, 204, 329
Lewis, C., 243
Liberal education, 241–42
Licensing of pets, 75, 233
Life expectancy of pets, 333–34
Lima State Hospital, 139, 160–61
Lippard, V., 250
Lithium chloride, 125–26
Lord of the Flies (Golding), 179
Lorenz, K., 77, 79, 152, 323, 331
Lycaon pictus, 13
Lynch, J., 43, 116, 153

M

McComisky, J., 140, 185
McCulloch, M., 142, 222

Mahler, M., 83, 85
Man Meets Dog (Lorenz), 79
Mann, T., 78
Marriage, disputes over pets in, 328
Maslow, A., 33
Massachusetts Society for the Prevention of Cruelty to Animals, 137–38
Massillon State Hospital, 156
Masuda, M., 149
Maturational Processes and the Facilitating Environment, 83
Mechanistic philosophy, 42
Medicine
    differentiated from veterinary medicine, 219
    education for, 250.
    (*See also* Therapeutic value and function of pets)
"Medicine, Biology, and Ethics," (Veatch), 250
Meggit, M. J., 8, 11
Meislick, D., 210
Mental illness, pet therapy for, 135
Mental retardation, pet therapy for, 159–60, 162
Mental Retardation Commission, 136
Mesolithic period, 5, 9–10
Middle Ages, 26
Minnesota Department of Health, 142
Miss Piggy, 244
Mongrels, 308
Montagu, A., 152
Morris, D., 65, 78
Mother-child bond, 81–82, 205
Motherese, 53–54
Mourning, 339–42. (*See also* Death of a pet)
Mugford, R., 140, 185
Multiple pet owners, 321
Mutualism, between pets and owners, 33–34
Myocardial infarction, pet therapy for, 102

N

National Institute of Mental Health, 137
Natufian, 151
Naturalism, in pet ownership, 34
Need-dependency relationships, 31–32
Neighborhoods, and dogs, 68–71

Neurological behavior problems, 304
Neurotic pet owners, 32, 321–22
Neutering of pets, 35, 37, 225, 306–308
Newberry Junior College, 138
New Eden: Animal Rights and Human Liberation (Fox), 29
New Horizons for Veterinary Medicine, 251
New man, J., 241–42
New York City dog laws, 233–35
New York State Psychological Association, 130
Nocturnal rodents, 16
Nonpet owners, 91–92, 94
Nonverbal communication, 150–51
Normal grieving, 339–40
Nursing homes, 147–48, 156–59, 256–57, 259

O

Obedience, 285, 305. (*See also* Dominance)
Object-oriented relationships, 30–31
Ohio State University, 142, 156–57
O'Leary Corson, E., 127
"One-man dog," 80–81, 135
Overattachment, 32
Overattention, 289
Overprotection of pets, 228–29
Owners of pets
    and aggression of pets, 229–30, 301
    and child abuse, 153
    as children to their pets, 79–81
    determinants of, 115
    education of, 238, 286
    and euthanasia, 275–83 passim, 336–37
    fears of veterinarians, 320–21
    grief, 339–42 (*see also* Death of a pet)
    hyperanxiety among, 222, 322–27
    idealization of pets by, 77–78
    laissez-faire philosophy of, 34
    neurosis in, 32, 321–22
    negligence of, 68–73 passim, 87
    numbers of, 68, 72
    obsessiveness of, 228–29
    personal characteristics of (general) 92–94
    potential problems of, 198, 203
    psychological benefits to, 114–115, 116, 198–99

seeking medical and social assistance from veterinarians, 327
sex of, 59–61, 282
submissiveness of, to dogs, 283–88

P

Paedomorphic characteristics of domestic dogs, 18–19
Papashvily, H. and G., 78
Parent-child attachment, 280, 282
Pavlov, I., 43
Peking Man, 7
Penal institutions, pet therapy in 160–61
People-Pet Partnership, 136, 257–59, 261–62
Perin, C., 199
Personality differences between pet owners and nonowners, 92–94
Petishism (Szasz), 65
Pet(s)
    abuse, 36, 162, 178–79, 215–16, 328
    and child development, 175–77, 328–29
    as child substitutes, 322–23
    empathy of, 56
    evocation of attachment by, 281
    as facilitators of therapy, 103–105, 118–20, 124–43, 154–62 passim, 168–72, 184–85
    grieving for, 339–42 (*see also* Death of a pet)
    life expectancy of, 333–34
    in marriage, 328
    matching, 119, 131
    as property, 335
    as significant others, 204–205
    as social assets, 48
    in urban areas, 35, 36, 231–38
Pet-A-Care Program, 137
Pets Are Good (Levinson and Freiberger), 142
Pets as a Social Phenomenon (PIAS), 193
Pet Food Institute, 72, 76
"Pets in Hospitals," (Cooper), 142
Phobias
    animal, 178, 227–28, 303–304
Physiological behavior problems, 304
Physiology and behavior course, 265–66
Play, as a benefit of pet ownership, 15, 115, 153

Population of pets, 68, 231
Postures, submissive and dominant, 287–89
Prayer, 57–58
Prevention of behavior problems, 283–93
Primate pets, 15–16
Proximity, and attachment, 272
Psychology
    of infancy, 44
    pathological biases in, 45–46
    practitioners, 131, 154
Psychosocial structure of nursing homes, 148
Psychotherapists, 130–31 (*see also* Pet facilitated of therapy)
Psychotics, therapy for, 102, 130
Pudd'nhead Wilson's Calendar, 156
Purcell, E., 250

Q

"Quality of life" and euthanasia, 336
Questionnaire
    to assess owner and pet behavior, 222–24
    for followup on dog behavior corrections, 316
Quigley, J., 253

R

Rabies, 233
Rahe, R., 149
Recreational uses of animals, 26
Relationships between owners and pets, 29–35, 47–52, 76–81 (*see also many other entries, including* Attachment; Attitudes; Children; Death of a pet; *etc.*)
Repression of sentimentality, 47
Research
    on blood pressure and speech, 54–56
    control of variables in, 42–44, 168
    on dog–child interactions in the home, 195–205
    on "one-man dog," 80
    on pet-facilitated therapy, 102, 105–119, 130–31, 139–41, 154, 156–68
    on physiological impact on attitude, 96
    on problem dogs and owners, 297–99, 300–305, 313

on the psychological determinants of keeping pets, 89–98
relating pet ownership to survival of illness, 47–52, 140–41
on tactile communication between people and dogs, 58–62
on urban pet planning, 238
on veterinary social work, 210
Responsibility, 241, 243–45
Retirement, 146–47
Riding for the Handicapped, 248, 50
Risk in pet-human interactions, 129
Ritual uses of animals, 26
Robb, S., 139
Rodents, 6, 16
Rogers, C., 56–57
Romasco, 210, 212–13
Rorschach ink blot test, 189
Rosenman, R., 117
Rynearson, E., 177, 178, 205

S

San Francisco Society for the Prevention of Cruelty to Animals, 137
Sanitation, 92, 143, 162–63
Sauer, C., 8
Scavengers, dogs as, 9
Scientific method, 42–44, 168, 242, 246, 250 (*see also* Research)
Scooper laws, 233–35
Scott, J. P., 79–80, 152
Sealth, Chief, 334
Searles, H. F., 65
Selective breeding, 18–19, 35
Self-esteem, 115, 147
Selye, H., 150
Sensitivity, 241, 243–45
Sentimentality, 46–47
"Separation anxiety," 84–85, 289–91, 303
Separation-individuation process, 83–84
Service of Prayers for Animals, 340
Sex
and affection, 44
of dogs, and their behavior problems, 289, 306–308
of owners, and attachment to pets, 282
and pet ownership, 36–37

and tactile communication with animals, 59–60
Shaw, G. B., 77
Shore, H., 149
Silver foxes, 19
Size of pets, 17, 341
Snoopy, 176, 244
Social
Darwinism, 44
environment, 92–93, 95
hierarchy (*see* Dominance)
interaction with pets present, 161–62
isolation, 47–50, 117
lubrication, 130
work, veterinary, 209–220
Socialization of canines, 80
"Social Sciences and Ethics," 250
"Song of Myself," (Whitman), 149–50
Speck, R. V., 195
Speech, 52–58
Spitz, R., 44
Standing postures, 288
Steinberg, S., 210
Stern, M., 117
Stewardship of animals, 34–35
Stock chasing, 306
Strays, 310–11
Stress, 150
"Study of PFT Intervention for the Impaired Elderly" (Robb), 139
"Study of the Use of Companion aids for the Physically Handicapped" (Willard), 140
Submissive postures, 287–89
Suicide, among the elderly, 117
Surgical modification of animals, 35, 37
Szasz, K., 151
Szasz, R., 65

T

Tactile communication with animals, 58–62
Talking to animals, 52–53
Technological era, 27–29
Television advertising, 151
Temperament of pets, 285, 308–310
Terminally ill patients, and pet therapy, 160
Territorial barking, 7, 11, 19

"Therapeutic Impact of Pet Animals with the Hospitalized Elderly," (Brickel), 139
"Therapeutic Role of Cat Mascots with a Hospital-based Geriatric Population—A Staff Survey" (Brickel), 140
Therapeutic value and functions of pets, 32, 41, 72, 102–105, 118–20, 124–43, 154–62 passim, 168–72, 184–85
Three Men in a Boat (Jerome), 156
Thurber, J., 78
Toilet problems. (*See* Elimination)
Touch, 58, 152–53
Touching (Montagu), 152
Toxocara canis, 233–34
Training for therapeutic assistance, 135–36, 138, 142
Trut, L. N., 19
Tufts University, 140
Type A Behavior and Your Heart (Friedman and Rosenman), 117

U

Understanding, 241, 243–45
University of Dundee, 195
University of Manitoba, 185
University of Michigan, Children's Psychiatric Hospital, 154
University of Pennsylvania, 253, 255
Unloving Care: The Nursing Home Tragedy (Vladeck), 147
United States
    Commission for Accreditation of Hospitals, 128
    pet population, 68, 151, 231
Urban pets, 36, 231–32

V

Valla, F., 151
Verbal communication, 52–58
Vermeulen, J., 160

Veterinarians
    openness of, to animal behavior specialists, 312
    owner's fears of, 320–21
    social work training of, 209–230 passim
Veterinary Medicine and the Study of Human Behavior (Ryder), 209
Veterinary School of the University of Pennsylvania, 137
Veteran's Administration, 128–29
    Medical Centers, 139
Videotape, in pet therapy, 158
Virchow, R., 121
Vladeck, B., 147
Voith, V., 296

W

Warwick, D., 250
Washington State University, 250, 253, 254, 256
White, E. B., 243
Whitman, W., 149–50
Wildlife management, 26
Willard, M., 140
Williams, A., 154
Wilson, W., 248
Wind in the Willows (Grahame), 243
Winnicott, P. W., 83
Wolves, 7, 10, 12–15 passim, 18

Y

Yoxall, M., 321–22

Z

Zemanek, M., 90
Zeuner, F., 6, 9
Zion, G., 243